SATPREM

A sailor and Breton,
Paris in 1923, Satprem took part in the French Resistance and was arrested by the Gestapo when he was twenty years old. Devastated after one and a half years in concentration camps, he journeyed to Upper Egypt, then to India, where he served in the French government of Pondicherry. There he discovered Sri Aurobindo and Mother. Deeply struck by their Message—"Man is a transitional being"—he resigned his post in the French Colonies and left for Guiana, where he spent a year in the middle of the jungle, then on to Brazil and Africa.

In 1953, at thirty, he returned to India to be near Mother, who sought the secret of the passage to the "next species," becoming her confidant and the witness of her experiences for almost twenty years. His first nonfiction work was dedicated to *Sri Aurobindo or The Adventure of Consciousness.* At the age of fifty, he brought out the fabulous logbook of Mother's own journey, *The Agenda*, in 13 volumes, then wrote a trilogy on Mother—*The Divine Materialism, The New Species, The Mutation of Death.*

At fifty-nine, he withdrew completely from public view to attempt the last Adventure: the search for the "great passage" leading to man's evolutionary future. In a last interview in 1984, *Life Without Death*, he described the beginning of his experience in the body, which he further developed in 1990 in *The Revolt of the Earth.*

THE MIND
of
THE CELLS

ALSO BY SATPREM

Sri Aurobindo or
The Adventure of Consciousness (1984)
By the Body of the Earth (1990)
Mother:
I. *The Divine Materialism* (1980)
II. *The New Species* (1983)
III. *The Mutation of Death* (1987)
On the Way to Supermanhood (1986)
Life without Death (1988)
My Burning Heart (1989)
The Revolt of the Earth (1990)

Mother's Agenda
1951-1973
13 Volumes

Recorded by Satprem in the course of numerous personal conversations with Mother, the complete logbook of her fabulous exploration in the cellular consciousness of the human body. Twenty-three years of experiences which parallel some of the most recent theories of modern physics. Perhaps the key to man's passage to the next species. (Vols. 1, 2, 3, 4, 5, 6, 7, 12 & 13 published in English)

Satprem

THE MIND *of* THE CELLS

Translated from the French by
Luc Venet

INSTITUTE FOR EVOLUTIONARY RESEARCH
1621 FREEWAY DRIVE, MT VERNON, WA 98273

This book was originally published in France under the title *Le mental des cellules* by Éditions Robert Laffont, S.A., Paris.

For information address:
U.S.A:
Institute for Evolutionary Research
1621 Freeway Drive, Suite 220
Mount Vernon, WA 98273

India & Asia:
Mira Aditi Centre
62 Sriranga, 1st Cross
4th stage, Kuvempunagar
Mysore 570023, India

Library of Congress Cataloging-in-Publication Data
Satprem, 1923-
the mind of the cells.
translation of: Le mental des cellules
1. Mother, 1878-1973 2. Spiritual life (Hinduism)
I. Title.
BL1175.M43S2613 1982 130 82-15659
ISBN 0-938710-06-0

Cover designed by Stephanie Baloghy
Manufactured in the United States of America
Second Edition. Fifth printing 1992

Beyond the tombs,
forward!

—Goethe

Contents

A PASSPORT TO WHERE?

Exactly fifteen days past my twentieth birthday, on a street corner of a French city, my life changed abruptly as I jumped to the sound of screeching tires and slamming car doors. Two men armed with revolvers leaped out of a *Kriminal Polizei* car, handcuffed me, and took me away. In thirty seconds it was over. Never again would I be an ordinary human being. The Gestapo, the interrogations under spotlights, night and day fading into each other, the footsteps of the SS men in the corridor at dawn. Would I be executed today or tomorrow? The frozen courtyards of Buchenwald, the rails embedded in the spotless tile floors of the showers—were they for a water ablution or a gas purge? And then, then. . . . The death of one man matters little. But the death of Man? The death of a child of man with all his dreams, his hopes, his belief in beauty, in love, in the immensity of a life that would be like a treasure to conquer, a continent to explore, a secret to discover. And then—and then NOTHING. Death, at least, is something. But nothing?

On that fifteenth day of November, of the thirty thousandth century since the advent of *Homo sapiens,* I found myself naked, devastated, as at the beginning, or the end, of time.

Man is dead? Long live Man! What is the meaning of a heart beating without its science, its gospels, its books—without a country, without law? Everything is dead, or not yet born. There is only that heart beating as if before the Flood, or after. There is only that young earthling gazing out, as if at the beginning of the world, over a vast empty beach where a seagull takes flight.

Indeed, what is the meaning of that heart without science, without knowledge, since all knowledge has crumbled, or is not yet born?

A heart beats for hope, for faith in the future. It sees the world as a great adventure to embark on—but what remains to be discovered when all the old goals are dead, when man's entire science is dead, when all the gods are dead, or not yet born?

It is terrifying. It is wonderful.

There are no more hopes, only the unknown Hope.

And I wonder if that child of man, who was then twenty years and fifteen days old, with a naked, empty heart, is not the forerunner of many children with similar hearts who will gaze out over the vast, empty beach of the world at the nullity of their science, of their bombs, of their devices—the terrifying and wonderful nullity of all the gods of the West and the East. And then, then. . . .

We are not at the end of a civilization.

We are at the Time when Man is to be born.

We have played enough with electric trains, penicillin, and electronic chromosomes. What if it is time for another game? What if it is time for a different kind of discovery, within a pure heartbeat? An unknown man under his worn-out coat?

At twenty-two, as I left that hell, I took Life—that deceitful witch—on my lap and said with fury, "Now, it's just you and me. You're going to tell me your secret, and no nonsense—

your secret that has nothing to do with any books, science or mechanisms; that is not from the West or the East, or from any country, but from the Country of the true Earth. Your secret that beats in my naked heart."

I moved heaven and earth. I tried everything. Oh, I wanted to force the secret out of this devastated flesh of man, of this empty, accursed and wonderful Earth. I roamed the continents; I listened to the shadow sounds of the gongs in Thebes and Luxor; I trod the red dirt roads of Afghanistan and unearthed Greco-Buddhist skulls; but there was still no smile on my lips. I climbed the slopes of the Himalayas, dug into eagles' nests in search of the Rajput treasure; I smoked opium like a drowning man, pounded at every door of this body; but the secret still eluded me. I dove into the jungle of Guiana, listened at night to the howlings of red monkeys like some primal animal chorus at the beginning of the world; I wandered through Brazil and Africa, searching for a mine of gold or mica, or anything; but the Mine within still would not divulge its secret. I returned to confront India; I practiced the secret of yogis, meditated with them, lost myself with them on rarefied peaks of consciousness; but the earth, *this* earth, still would not reveal its Marvel. I became an errant beggar; I wore out this body to the bone, praying in temples, knocking at every door; but the one Door that could soothe my heart would not open.

And again I was naked. Was there no hope, other than accumulating electronic gadgets and bombs and false wisdoms, or true ones that take you straight to heaven but still leave this earth to rot on two feet?

Now, I was thirty.

It was still the thirty thousandth century since the advent of man. Damn it, had all those millions of years gone by merely to end up walking around in a gray flannel suit with a briefcase in hand and a stamp on my passport?! A passport to *where*? A stamp for *what*? What had happened to Man as a great adventure, as a secret to discover, as an unknown treasure?

I was born in Paris. I could have been born in Tokyo or New York—but what about being born to the world? Being born, at least, to something other than a grandfather and a great grandfather and a high-school diploma and books lined up in dead libraries—the same old story that repeats itself endlessly, in English, in French, in Chinese; in a man, who dies again and again without ever finding what makes his heart beat, or why the flight of a seagull over a little beach suddenly fills him with such breath, as if he himself could fly?

My passport says I cannot fly, except in a Boeing 747.

But my heart says otherwise. And the heart of the whole earth is beginning to say otherwise.

When I was thirty, I met She who said otherwise. She was eighty years old then, and she was as young and mirthful as a young girl. She was called "Mother." It was in Pondicherry on the Bay of Bengal.

Mother is the most wonderful adventure I have ever known. She is the door that opens when all the others have closed on nothing. For nineteen years, she took me with her along untrodden paths leading to Man's future, or perhaps to his true beginning. My heart beat as it had never beaten before. Mother is the secret of the Earth. No, not a saint, a mystic, or a yogi; she is not from the East or the West; she is not a miracle worker either, or a guru, or the founder of a new religion. Mother is the discoverer of the secret of Man when he has been shorn of his gadgets and his religions, his spiritualisms and his materialisms, his ideologies of the East and the West—when he is himself, simply a heart beating and calling out to the Truth of the Earth, a body calling out to the Truth of the body, as the cry of the seagull calls out to the wind and the open air.

It is her secret, her discovery, that I will try to tell you.

For Mother is a fairy tale within the cells of the body.

What is a human cell?

Another concentration camp—a biological one?

Or a passport to . . . where?

S.
July 8, 1980

INTRODUCTION

We are before an incredible mystery which could well be a fairy tale.

The fairy tale of our species.

Let us begin at the Galapagos Archipelago, where, around 1835, Darwin conceived his theory of evolution: iguanas are not forever iguanas, nor men forever men. We have heard nothing more serious since—or more captivating, or let us say liberating, for it is indeed a release from captivity that we are after. But what *kind* of release, if we leave aside blowing up the planet or seeking heavenly or other yogic forms of salvation which, we begin to realize, leave the planet unchanged?

"Salvation is physical," says Mother, whose adventure into the cells' consciousness we are about to relate. Evolution is obviously materialistic, or at any rate material. But what sort of matter is that? Is it closed or open? Darwin opened it, so did his contemporary Jules Verne. Max Planck, Heisenberg, Einstein, and their Impressionist, Fauvist and Pointillist friends opened it too—matter was exploding on all sides. This is also what Sri Aurobindo and Mother are about, along with a few modern physicists. So why should matter be less open with biologists?

When Darwin died in 1882, Sri Aurobindo was ten years old. He had already left India to learn the lesson of Western materialism in London. Mother, his future companion, was four years old in Paris, and Einstein was three in Ulm.

In 1953, another serious theory saw the light of day, but when the "serious" takes on the aspect of a prison we become suspicious. Because since the advent of vertebrates, four hundred million years ago, evolution has passed over a good number of "biologies" and "philosophies" one after another, including those of the crab, the rabbit, and the orangutan, in order to produce a next species. How that passage occurs is what interests us. In 1953, then, a team of Anglo-American biophysicists discovered the mechanism of duplication of DNA. This is very serious indeed. The sequence in the chain of amino acids defines for perpetuity whether we will be mouse or man. A particular, magical and perfectly scientific molecule, deoxyribonucleic acid, or DNA, inexorably controls that sequence from father to son—unless some X-ray or cosmic radiation (or a little bomb) should cause a derailment in the DNA sequence and give rise to a monstrosity rather than a new species. Only after thousands or millions of years of imperceptible "favorable" mutations in the DNA of germ cells (which would eventually add up to a major change) we should land in another species—provided the bomb does not explode in the interim, and the present five billion members of *Homo sapiens* do not produce enough billions of rat sapiens to decimate the earth. This is also a serious consideration, for it took thousands of years for the human population to reach the one billion mark—in 1830—but only one hundred years for the second billion, thirty years for the third, and a mere fourteen years for the fourth billion.[1]

---

1. *New York Times* of March 16, 1980.

The problem is urgent. We no longer have thousands of evolutionary years to solve it—perhaps not even ten. So how to find a way out, despite the Anglo-American team and the need for repetitive cellular mutations? Some modern evolutionists have theorized about "bursts of evolution," since the long process of Darwinian natural selection does not seem to account for major changes among species. But then, what triggers those "bursts"?

If the solution is not in heaven or in any yogic release, can it be found directly in the cells and in matter? That man will not indefinitely remain man, or even an "improved" man, is not in doubt, just as the reptile could not remain a reptile in the dried-up swamps of the Secondary Era. If we do not find the "key," evolution will find it for us, in spite of the biologists.

Seventy million years ago, the dinosaurs suddenly disappeared from the face of the earth they were devouring, to leave room for the gerbils and field mice to gambol and frolic.

> **57.412** — Can we expect this body, which is now our means of manifestation on earth, to be able to change progressively into something capable of expressing a higher life, or will we have to abandon this form altogether and enter another form still unknown on earth? asked Mother, who was precisely trying to find the "key" for the species in the cells of her body. Will there be continuity or a sudden appearance of something new? . . . Will the human species be like certain other species that have disappeared from the earth?

This was in 1957.

Darwin took more than twenty years to dare to write what he had first sensed in the Galapagos: *The Origin of Species* was published in 1859. And then again, he said it was "like confessing to a murder." Before Mother's story, I feel a little the way Darwin felt about his iguanas: "Come now, is that possible?" What will Biology say, and what will Medical Science say, and . . . Yet, there cannot be any doubt. For nineteen years, I listened to Mother describe her experiences, as she continued Sri Aurobindo's work. At the time, I did not fully grasp the significance of her words. And then, one day in 1973, at the age of 95, she left, leaving me dazed before a mountain of documents filled with meaning and incomprehensible at the same time. For seven years I have explored these documents, struggling with them, banging my fists against the wall and calling to Mother from beyond that "stupid death," as she called it, to help me decipher her secret—yet it is there, completely open in the thousands of pages of the *Agenda*. But what sense does the experience of a mouse make to a dinosaur? Yet, it is full of sense, it's right there, but a little something has to click for all the pieces of the puzzle to fall into place. I even wrote three volumes [1] to attempt to follow the thread and chart a path into this incomprehensible future of man. Oh, how I agonized! Sometimes, I even felt like Conan Doyle's Sherlock Holmes with magnifying glass and mental logic trying to grasp something that has ceased to be mental. Mother is a frustrating, fascinating mystery thriller about the history of the future species. How is a future species made? How does it emerge? And by what means? Through what mechanism? Someday it all becomes obvious—but nothing is more invisible than the obvious, because it is so close to us that we don't see it. Can mice or even monkeys see something in men? They must think our tree-climbing leaves something to be desired, but beyond that? So I scrutinized and analyzed Mother's story, but beyond that? I stared wide-eyed at it—and yes, it is a little like "confessing

1. *Mother*, Vol. I, *The Divine Materialism*, Vol. II, *The New Species*, Vol. III, *The Mutation of Death* (New York: Institute for Evolutionary Research).

to a murder." How well I understand what Darwin meant. This is such a challenge to our species and its laws, and yet it is logical, natural. But try to tell a mouse that *Homo sapiens* is natural and logical!

The only way to introduce the reader to this biology-mystery thriller of the future species is to set forth Mother's decisive experiences directly, as laboratory experiments, without flourish or comment. We will then follow the lines that led to those experiences, then move on to those that developed into a new nucleus of experiences, until the puzzle is complete and the conclusion inevitable.

We are not concerned here with any kind of mysticism or philosophy, even Hindu, or with any scientific explanation, for what is the reptile's science to the archaeopteryx? We are concerned only with the facts of the experiment, however strange they may seem. And, like Darwin in the Galapagos, we begin with a basic statement no evolutionist will deny. Mother's first premise:

> **58.2811** — The physical substance evolves through each individual formation, and one day, it will be capable of bridging the gap between the physical life as we know it and the supramental life that will manifest.

The body is the bridge.

That is, the cells.

The cells behaving according to the Anglo-American scenario, or otherwise?

Countless imperceptible mutations over a period of thousands of years, or an abrupt change: "The miracle of the earth," as she would say, the fairy tale of the species?

Though a perfectly biological and earthbound fairy tale.

58.145 — It seems one can truly understand only when one understands with the body.

54.214 — Knowledge, for the body, is to be able to *do.*[1]

In truth, Mother is the most incredible revolution ever carried out by man since the Pleistocene day when the first hominid began counting the stars and his sorrows.

1. The first two digits at the beginning of each quotation refer to the year of the particular experience as noted in *Mother's Agenda* (New York: Institute for Evolutionary Research).

MOTHER, otherwise known as Mirra Alfassa, was born in Paris in 1878, of an Egyptian mother and a Turkish father. She was a year older than Einstein, and a contemporary of Anatole France, with whom she shared a sense of gentle irony. This was the century of "positivism"; her father and mother were "all-out materialists," he a banker and a first-rate mathematician, she a disciple of Marx until the age of eighty-eight.

Yet when she was very young Mirra had strange experiences involving past history and perhaps the future; she met Sri Aurobindo "in a dream" ten years before going to Pondicherry and took him for "a Hindu God dressed in the garb of a vision." Equally at ease with higher mathematics, in front of an easel, or sitting at a piano, she befriended Gustave Moreau, Rodin and Monet. She married a painter, whom she later divorced to marry a philosopher who took her to Japan and China at the time Mao Tse-Tung was writing his first political essays, and to Pondicherry, where she met Sri Aurobindo, with whom she stayed thereafter. She spent thirty years beside he who, at the turn of the century, was announcing "the new evolution": "Man is a transitional being." After Sri Aurobindo's death in 1950, left in charge of a huge ashram that seemed to represent all the human resistances of the world, she plunged into the "yoga of the cells" and finally discovered "the great passage" to another species. Isolated, misunderstood, and surrounded by human resistance and ill will, she left her body in 1973 at the age of ninety-five.

"I don't think there was ever anyone more materialistic than I, with all the practical common sense and positivism," she would tell me in the midst of her dangerous experiences in the consciousness of the cells, "and now I understand why I was that way! It gave my body a wonderful sense of balance. All the explanations I sought were always of a material nature; it seemed so obvious to me: no need for mysteries or thing of that sort—you must explain things in material terms. Therefore I am sure there is no tendency for mystical dreaming in me! This body had nothing in the least mystical in it, thank God!"

1
THE NEW ELEMENT

There has been a turning point in the development of our species, but it was probably preceded by many small, sporadic and unintelligible breakthroughs labelled with one name or another—indeed, who could recognize a breakthrough into another species? It is only *after* we have fully become a man that we can say, "Aha, so this is what a man is." And we say it only after many gradual experiences have reassured us that we are not delirious monkeys or sick and decadent primates, for the initial reality of a new species is everything it misses from the old: man's qualities are the monkey's weaknesses.

Hence, breakthroughs into "another," out-of-the-ordinary condition have more than likely taken place microscopically, at different physiological levels, over hundreds and thousands of preparatory years, but always in total ignorance that they were breakthroughs into "the other state." Before the small Borneo lemur acquired binocular vision, which prepared ours, a few species must have experienced a number of disconcerting "visions," which were nonetheless "logical," "mathematical," and natural for this fish or that bat. And even now, what is our human retinal vision but a narrow light-spectrum between ultraviolet and near-infrared? Also, because the evolutionary breakthrough always reverts to the former state until the decisive appearance of the new species, it is inevitably expressed in the language and customs of the involuntary subject undergoing the experiment, and hence covered in a heavy cloak that distorts almost completely what could have been the pure experience of the other state. Throughout time and in every conceivable language, we have seen many "mystics," "madmen," and "hallucinators," and we have tended to accept or even glorify those who best complied with our own notion of Good, Beauty, Apocalypse, or Heaven. But what is a bat's notion of beauty to a wren? The bat just feels a little "dazzled." Yet "something" was there, if only the heaven of a mystical bat.

For *Homo sapiens*, these breakthroughs have taken place at various levels of his being. And since he is shut in a mental shell, as the sea urchin was shut in its calcareous carapace, the rock in its cloak of electrons, and the monkey in its vital strength, it is primarily at the mental level that attempts to break through had to occur most often: one loses consciousness on the operating table or in a mystical trance or simply during sleep, and one emerges in another world. A certain cancellation of the old system seems necessary in order to gain access to "elsewhere." And it makes sense: one cannot go to the next species' "heaven" in human boots, any more than the archaeopteryx can take off on its inaugural flight with the body of a reptile! As we said earlier, it is the weaknesses of the old species that open the door to the next one—but a door must open. Throughout the ages, therefore, we have opened many doors in our heads, and less frequently in our hearts; we have even gone down the physiological ladder and opened the gates of the lower abdomen, letting all sorts of hellish or cruel little entities take control of us: many kinds of rabid, subhuman creatures that are still quite abundant on this earth; not to mention those who have left the species altogether, from above, in a nirvanic or ecstatic rocket, and who have sometimes borne strange and rapturous mumblings. For poetry, too, is a "translation" of that elusive other state that our species is yearning to grasp but does not know how. How does one grasp the next species? It is neither at the mental, nor cardiac, nor umbilical, nor even pelvic level that one goes into the other state, into "the thing," as Mother used to say, for she found no other word for it. To be more precise—for one cannot say dogmatically and categorically that all those breakthroughs were meaningless—it isn't at any of these levels that one reaches the pure "thing," without distortion, *in its original language*. The next species must come into being in the body. It is obvious. Unless it is experienced in the body, at the physiological or cellular level, it will remain but a translation into a foreign language, through layers of sleep, ecstasy, or meditation, which may produce all kinds of fabulous and amazing things, but refracted rays nevertheless, the translation of "something," perhaps like what the goldfish sees of man *through* the glass of its fishbowl. We do not know whether we look angelic or diabolic from its side of the glass, but we are certainly "something happening."

But if we talk of a breakthrough at the "cellular level," biology will immediately jump on us with its inexorable and imperturbable chain of amino acids from father to son, except for a few pathological variations. "How do you intend to change the sequence of nucleotides in the DNA to produce your next species? Will it have fins, wings, or a third eye?" At a certain stage in evolution, it must have been very difficult indeed for a manganese nodule to conceive of a brazen, moving flagellate organism. A new species is always impertinent in the eyes of the old. And yet, there must be a link between the two, a connection to bridge the gap—a handle. Our deficiency is not only a lack of imagination about the future, but above all an incapacity to conceive of anything but an improvement or an extension of the present. Our future man would still be a man plus this plus that and minus this minus that. Is a rhizopod an extension of manganese? Is man an extension of the fern? Perhaps he is something entirely different. But how does one connect with this "entirely different something"? We have no idea of what will establish the connection because we don't know where the other side is. Yet it is in the body.

In other words, the next species may be another *kingdom* altogether, as different as the field mouse is to the fern. Not a man plus, but another being, another life form in matter, after the mineral, vegetable and animal kingdom, of which we are a part. Still, there must be a connection with it, just as the virus is the bridge between life and inanimate matter. What is the bridge with "overlife," to use one of Mother's halting expressions? And what is the nature of that life? Science's view may be right that the accumulation of mutations, the modifications of DNA in the germ cells, is what creates a new species in the course of time, but what triggers those mutations? "[Mutations] seem to us in our ignorance to arise spontaneously," acknowledged Darwin in his time. And our "ignorance" is *not* dispelled by our modern knowledge of DNA; it is only clothed in scientific language. Today's science assigns the following as "natural" causes of mutations: 1) random errors in the duplicating of DNA when a cell divides into two daughter cells, and 2) cosmic rays. In other words, chance, chance, chance.

But have we thought about the creature undergoing that mutation? What does she, he, or it have to say? Could it be that, at least partially, it *wills* its own mutation; it aspires for a change of air because of its own suffocation (or growing sense of inadequacy) in the milieu? It is conceivable that the species itself, or some pioneers within it, participate in the evolutionary push, handle and direct the evolutionary force consciously—collaborate with it—and let it fashion a new physical way of being until a new equilibrium is reached, a more satisfying poise in the milieu? A "collaboration" that would mean a world of difference in the swiftness of the process! A "burst" of evolution? That is to say, what we call mutation may be only the external result of an inner pressure initiated by the creature itself for reasons of its own, a visible consequence whose cause has so far escaped our electron microscopes and carbon-14 investigations. (There exists in fact a need for a new scientific approach that would consider the part played by the creature in its own evolution, that would begin to look at evolution as more than a one-sided—environmental—affair and welcome the admission of the second player.)[1]

But undoubtedly, for a new species to come into being, there must be some modification within it, a new element. What is the nature of the change in the fern with respect to the mineral, and that of the animal with respect to the vegetable? We are obsessed by forms—the form—but what has changed from one kingdom to another except the rate of movement? There has been a transition from the inertia of the rock to the accelerated growth of vegetable life, and another to the dynamic explosion of animal life—all transitions in the rate of movement. Now physicists will stir and tell us about electromagnetic waves and the whirl of electrons around the nucleus. Einstein told us about relativity: the parameters of a physical event are related to the speed of the frame of reference

1. In September 1988, *The Boston Globe* reported the following: "In a provocative challenge to accepted evolutionary theory, two groups of scientists are asserting that simple organisms can reengineer their own genes in response to environmental stress. New experiments show that even single-celled organisms can profit from experience and *choose which mutations they should produce.* 'It seems that bacteria are doing something one would have thought impossible,' acknowledged Dr. John Cairns, the leader of a Harvard School of Public Health group of researchers. 'My guess is that if cells have evolved mechanisms that allow them to do this, it is going to be so advantageous that it seems unlikely that higher organisms will have given that up.' "

in which this event takes place. To say things simply, speed has to do with distance, distance has to do with the six legs of an ant, the two wings of a seagull, the two feet of a man, or even a jet plane; but all these are just faster- or slower-moving animals equipped with a more or less ingenious machinery to join with what is "far away" or "outside." But it is possible that the next "machinery" or "organ" of the next species will result in an even greater acceleration, as it were, to abolish the sense of "outside" and "far away" and to make the "distance" of a flagellum or a jet plane as outmoded as the rock's inertia is to a living being. What is that machinery or "organ" capable of giving us such rapid movement we could reach the farthest galaxies in an instant and eliminate distances as if everything were taking place within us, and yet in a body of cellular, earthbound matter? Is there a functioning in the body that would allow us to be simultaneously in New York, Borneo, or God knows where while remaining contained in the layers of cellular walls that make us a man rather than a mouse? If we were physiologically, or "geographically," granted such "supernatural" movement, this would undoubtedly constitute another species, another kingdom. What is "natural" for man may be supernatural for a fish, for the notion of what is natural no doubt evolves from one species to the next, and "The supernatural is that the nature of which we have not yet attained," as Sri Aurobindo said.[1]

But then, where in the body would that curious new functioning be located—a functioning that would not cancel our precious germ cells but would give all the cells of our body a brand-new mode of being, perhaps even a new geography seen with non-binocular eyes? What would become of the jet plane and all our damned machinery, including the telephone and

1. *Thoughts and Aphorisms,* XVII, 88.

the space shuttle? This is indeed an entirely different notion of space and time—a different "frame of reference," a different determinism—and it may be as disconcerting as going from the rock's peaceful inertia to the swarming of the vertebrates. And what would become of death, then? What would become of matter in that new "system"? What is matter—its electrons, its cells and galaxies—seen by a non-binocular organ, or through something other than a microscope or telescope, which are but an extension of the same archaic retinal vision?

Biology and physics describe the laws of a given environment, of a human fishbowl that tries to observe itself through the glass walls of the bowl, but when the environment changes, as it did when the amphibian emerged into the wide-open air of life, all the old laws crumble and another unpredictable "life," or "overlife," appears.

The "link," however, remains to be found. If it is not found in nirvanic pirouettes and ecstasies, in mental convolutions, or in the sleep and dreams of this painful human species (which may have been conceived to live a real paradise on earth, in a real body without death and confining walls), where on earth is it to be found? From one species to another, from one kingdom to another, we have gone from one narrow prison to another not so spacious one. Could it be that the next kingdom will be that of a spacious and unshackled man?

With Mother, for the first time ever, the creature is caught (and recorded) in the act of evolution. Mother is the story of a willed mutation of our species. Instead of fleeing to mystical heights and poetry, we will join her adventure down into the consciousness of the cells in search of the next environment, the *new element* that will open the doors of our prison and cast us onto a new earth, just as, one day, the first amphibian landed on the sunny shores of a new world.

57.107 — A new world is BORN. This is not the old world being transformed; this is a NEW world being born. And we are right in the midst of the transition period when the two overlap, when the old still persists, all-powerful and entirely in control of the ordinary human consciousness, but the new slips in, still quite modest and unnoticed—so unnoticed that externally it hardly disturbs anything—for the time being—and is even absolutely imperceptible to the consciousness of most people. Yet it is working, it is growing.

56.103 — Every time a new element is introduced among existing combinations, it causes what could be called a tearing apart of the limits. . . . The perceptions of modern science undoubtedly come much closer to expressing the new reality than, say, those of the Stone Age. But even they will become suddenly outdated, surpassed, and probably shattered by the introduction of something that did not exist in the universe as we have studied it. It is that change, that sudden alteration of the universal element which will probably cause a sort of chaos in our perceptions, from which a new knowledge will emerge.

That "new element" is the mind of the cells, which is in the process of disrupting our human world just as our thinking mind disrupted the world of the apes.

2

THE OTHER STATE

A first experiment always seems strange, even a little wild. Yet there must have been a day when for the first time on this planet one last old reptile became the first young bird. How does it feel when you suddenly take off, and there has never, ever been a bird before in any logical and reasonable reptile sky? It is quite out of place, and many an old dinosaur must have shrugged their dorsal spikes: "This isn't possible, it must be an hallucination." From one hallucination to another, we have become little men in gray flannel suits. And now, what's next?

One morning in January 1962, I saw Mother arrive a little pale, but with her customary self-mocking air, as if irony were the only bearable way to handle the new species without derailing from the old. She was eighty-four years old. In her quiet and amused tone, she said:

> **62.91** — It's an odd thing, there are these strange attacks that seem to have nothing to do with my health. It's like a decentralization. You see, in order to form a body, the cells are concentrated and held together by a kind of centripetal force; well, this is just the opposite action! It's as if the cells were being spread out by a centrifugal force. And when it becomes a little too much to bear, I leave my body, and the outer, apparent result is fainting; I don't actually "faint," because I remain fully conscious. So it does create a bizarre sort of . . . disorganization. . . .

At first, the new species obviously means the disorganization of the old.

> The last time it happened, someone was there, so I didn't fall and hurt myself, but this time I was alone in my bathroom and . . . evidently experiencing a movement of consciousness in which I was spreading myself out over the world—spreading myself PHYSICALLY. That's what is so curious, it is a sensation of the cells! I was experiencing an increasingly intense and rapid movement of diffusion when, suddenly, I found myself on the floor.

The experiment follows a specific course. We will first outline that course before describing in detail how and through what process and transitions Mother reached that particular point. The fact is, she left a certain human state to enter another state or another environment, the way the amphibian had. The description of this new environment will help us to better understand the old one and what caused the barrier between the two. That barrier is really the crux of our problem; it must be located at cellular level since this is where the breakthrough occurs.

> **62.155** — For example, every day I walk a little to reaccustom my body (I walk with someone's help), and I have noticed a rather peculiar condition, something I could describe as giving me the illusion of a body! I entrust my body to the person helping me, (it's no longer my responsibility, you see, it's that person who has to make sure it doesn't fall or bump into anything), and the consciousness is like a consciousness without limits, which feels like waves, but not separate waves; it's a whole MOVEMENT of waves, a movement of material, corporeal waves, as it were, as vast as the earth and not . . . round or flat—something that feels very infinite and undulating. And this undulation is the movement of life.

We are now in modern physics! It is a fact that all physical theories attempting to explain the structure of our universe and the composition of matter use the wavelike or sinusoidal movement as the fundamental constituent and dynamic foundation of physical reality. Whether in electromagnetic, gravitational fields, or atomic interactions, in the heart of the atom or at the confines of our universe, everything moves and propagates as "waves." "This undulation is the movement of life," says Mother in a striking way. And she adds:

> And the consciousness (of the body, I suppose) floats in this with a sensation of eternal peace. But there is no sense of size whatsoever; it is a movement without any limit, with a very quiet, vast, and harmonious rhythm. And this movement is life itself. I walk about my room, and that's what is walking. It is very silent, like a movement of waves with no beginning and no end, with a condensation like this *(vertical gesture)*, and a condensation like that *(horizontal gesture)*. And it moves by expansion *(gesture of a pulsating ocean)*. That is, it contracts and concentrates, then expands and spreads out.

Amazingly, this description reminds us of the electromagnetic field with its two perpendicular components, the electric field and the magnetic field, and of their propagation along an infinite sinusoidal wave. Here, we touch on a stupendous mystery: how is it possible for a material, cellular body to *be* a wave that carries the worlds in its infinite movement and governs the existence of atoms and galaxies? How is it possible to *be* an infinite, omnipresent electromagnetic wave while remaining within the narrow confines of a human body, which still faints a little at the beginning because it is not accustomed to its new condition? In other words, a body the size of the universe.

The experiment would continue eleven more years, gradually becoming more precise and "familiar," but with an often puzzling vocabulary, because Mother sometimes used one word, sometimes another, giving the impression of different phenomena and especially different worlds, while she was all the time describing the same process in the same material world. But try to describe matter as seen through the eyes of a bird to a stubborn goldfish that perceives everything through the walls of its fishbowl! To the goldfish, that description isn't of real, solid matter. It may even seem a bit supernatural. What "words" could Mother have possibly used to describe something that is still nameless? "Electromagnetic wave" comes *later.* At the time, it is only "something happening."

Mother's first cry following her exposure to the total experience, which occurred in April 1962, takes us by surprise:

> **62.134**—Death is an illusion, sickness is an illusion, ignorance is an illusion!—something that has no reality, no existence. . . . Only Love and Love and Love—immense, formidable, stupendous, carrying everything. And the thing is DONE.

The transition to the next species is done. Once the first bird has flown among the reptiles, others will inevitably follow. But the essential point is that death and sickness *materially* disappear in that other state; since Mother's experience is a bodily one, felt and lived by the cells, not a mystical experience on nirvanic heights. This has nothing to do with "the world as an illusion" preached by the mystics; the illusion is about our physical perception of the world and its resulting falsehood: illness and death. When cellular perception changes, illness and death vanish and change into . . . something Mother was gradually to discover.

The experiment continues:

> **62.121** — I am constantly faced with a problem—quite a concrete and material problem when dealing with cells that must

remain cells and not dissolve into a nonphysical reality, while at the same time being supple, flexible enough to expand indefinitely. This body is indeed very difficult; it is very difficult to keep it from losing—how can I say?—its center of coagulation and dissolving into surrounding matter.

61.252 — This body no longer functions as usual. It is scarcely more than a center of concentration, like an aggregate of something; it isn't a skin-bound body at all. It's like an aggregate, a concentration of vibrations. And even what is usually called an "illness," or even a functional disorder, does not have the same meaning for this body as it does for the doctor or layman, for example; it isn't like that, the body doesn't feel it like that. It feels it as a sort of . . . difficulty in adjusting to a new vibratory need.

62.185 — The only sensation that remains in the old way is physical pain. I have the impression that these are the symbolic points of what remains of the old consciousness—pain. Only pain feels the same as before. For instance, food, taste, smell, vision, sound are all completely changed. They belong to another rhythm. That is, the organs function differently—is it the organs themselves or only their functioning that have changed? I don't know, but they obey a different law. The only thing that remains materially concrete in this world—this world of illusion—is pain. It seems to me to be the very essence of Falsehood. I am even forbidden to use my knowledge, my power, my force to cancel pain as I used to do—I did it very well before. It's strictly forbidden now. But I have seen that something else is intended, something which is in the making. . . . It is still—I can't say a miracle because it isn't a miracle—but it is a wonder, the unknown. When will it come? How will it come? I don't know.

Indeed, the question was no longer that of abolishing pain or suspending death by means of higher "powers," yogic or otherwise, but of *transforming* pain and death through the natural power of the cells themselves. This is the whole "Yoga of the Cells." The next species has nothing to do with extraordinary new organs or stupendous powers, but with a new cellular functioning and perception, which will radically and *naturally* change the condition of our present, death-ridden bodies.

> **62.315** — Now, I make a constant distinction between . . . (what shall I say?) life in a straight or angular line and undulating life. There is a life in which everything is sharp, hard, angular, and you bump into everything; and then there is an undulatory life, very soft, very charming—very charming—but not too, too solid. It's strange. It is a completely different kind of life. Even goodwill is aggressive, even affection, tenderness, relationships—all that is so very aggressive. It's like being beaten with a stick. Whereas "that" . . . is a sort of cadence, a wavy movement of such breadth, such power! It's simply fantastic. And it disturbs nothing, moves nothing, clashes with nothing. And it carries the universe in its undulatory movement so smoothly!

Could this have something to do with Einstein's famous "unified field"?

> **68.32** — In practice, if something goes wrong for one reason or another (a pain or a disorder somewhere in the body), with "that," the disorder disappears almost instantly. And if I remain patiently in that state, even the MEMORY of the disorder disappears. That's how the disorders that were recurring like habits gradually disappear forever.

> **68.1610** — The consciousness has curiously become increasingly intense and spread out, and the body is like something floating passively in that consciousness. I don't know how to explain it. It's like an ocean of light continuously at work, with something floating in it. . . . It's dark aquamarine. You know the color?

> **68.32** — But only when it is ready will the body be able to let itself go in that way. And this is now what is being prepared. The movement is, yes, to melt completely, with the result being the annihilation of the ego—in other words, an unknown state, or,

rather, one not realized physically; for all those who sought Nirvana did so by abandoning their body, while our work is to make the body, the material substance capable of melting. This is what we are trying to do. How to keep a form without an ego is the problem. That's why the work is being done very slowly and gradually. That's why it takes time: each element is taken in turn, then transformed. The wonder is to keep a form while completely shedding the ego (for the ordinary consciousness, it's nothing short of a miracle). In the vital [life-nature] and in the mind, it is easy enough to comprehend, but HERE, in the body. . . . How do you prevent the body from dissolving in that movement of fusion? Well, that's precisely the experiment underway, the fascinating course being followed at the moment. There are times when you feel that absolutely everything is disintegrating, falling apart, and at the beginning, when the physical consciousness was insufficiently enlightened, it felt like: "Oh, these must be the first signs of death!" But then, gradually, came the knowledge that it was not the case at all; it was only the inner preparation to strengthen the aptitude. And I clearly saw that, quite the contrary, once this very particular sort of plasticity, this extraordinary flexibility was achieved, it would naturally lead to abolishing the need for death. You see, every time the rule or domination of ordinary laws is replaced at a given point by the other authority [that of the other state], it causes a transitory condition that has all the appearances of a tremendous disorder and a very great danger. And as long as the body does not know, as long as it is still in its state of ignorance, it panics and thinks this is a serious illness; but that's not what it is originally: it is the removal of the ordinary law of nature and its replacement by the other. Therefore, for a moment, things are neither here nor there, and that moment is critical.

69.164 — The body is strangely fragile at the same time. It feels as if it had left all the ordinary laws behind and were just . . . suspended in midair. Something is trying to establish itself. And it is extremely sensitive to anything it comes in contact with. It works both ways: extremely sensitive to what comes from others, and at the same time with an extraordinary power to enter them and work in them. It's as if a certain type of boundary had been eliminated.

62.275 — It is a very impersonal sort of condition in which all the habits of reacting to external things around you have completely disappeared. But it hasn't been replaced by anything. It's . . . a wave. That's all. When will it become something else? I don't know. And you can't try to make it happen! You can't make any effort, or seek anything, because it immediately causes an intellectual activity which has nothing to do with "that." That's why I conclude it is something one must become, live, be—but how, through what method? I have no idea.

How can a fish try to be anything *other* than a fish? Even its efforts and ideas will be those of a fish.

62.66 — For the ordinary vision, externally, superficially, there seems to have been a great deterioration; yet the body doesn't feel that way at all! All it feels is that a particular movement, effort, or gesture—a particular action—still belongs to this world, this world of Ignorance; it is not done in the true way; it isn't the true movement. And it senses, or perceives, that the condition I described earlier as gentle, smooth, without angles, must develop along specific lines and produce bodily effects, enabling the true action. There is a new way to be found. But not "found" like that, in one's head; that way is being ELABORATED somewhere. Things have become so acute that whenever I change states, suddenly I feel as if my body were encased in rough files and wood splinters, while in fact it is comfortably seated on feather pillows!

In that exploded space, the sense of time,too, changes. One morning, Mother said to me, laughing:

62.147 — One day, we will say, "Do you recall how we thought we were doing something in such and such year! . . ." Believe it or not, I suddenly found myself projected into the future: "Do you remember, back then?" (It's always located to my left; I wonder why.) "Do you remember, back then? How we thought we were doing something, and knew something! . . ." It was quite funny. You see, in the ordinary consciousness, there is a sort of axis, and everything revolves around that axis. This axis is fixed somewhere, and everything revolves around it—that's the ordinary consciousness. And if it moves ever so slightly, we feel lost. It's like a tall axis—it varies in size; it can be quite small—planted squarely in time, and everything revolves around it. It may extend more or less far, be more or less high, more or less strong, but it all revolves about that axis. And for me, now, there is no more axis. This is what I was looking at—it's gone, finished, blown away! It can move here or there *(gesture to the various cardinal points)*, it can go backwards or forwards; it can go anywhere. The axis is gone; things no longer revolve around an axis. It's interesting. No more axis!

But suddenly, the "undulation" becomes very concrete and reveals its true nature: the constituent and dynamic foundation of all physical reality.

63.108 — There must be something new in the consciousness of cellular aggregates . . . something, a new experience must be in progress. As a result, last night I had a series of fantastic experiences—cellular experiences—which I don't even understand and which must be the beginning of a new revelation. . . . As the experience began, something in me was watching (there is always something ironical and amused in me witnessing everything), and it said: "Well, if this were happening to anybody else, he would think he is seriously ill or half mad!" So I stayed very quiet and said, "All right, let it be. Let's see what happens—I'll just see what happens. . . ." Indescribable! Indescribable! (It will have to be repeated several times before I begin to understand.) Fantastic! It began at eight-thirty and lasted till two-thirty in the morning, which means I did not lose consciousness for a second, while I lay there, witnessing the most extraordinary things. I don't know what will happen next. . . . It's indescribable! You see, you become a forest, a river, a mountain, a house—and this is a sensation of

THE BODY, a perfectly concrete sensation of the body. And many other similar things. Indescribable.

(Question:) A kind of ubiquity of the cells?

Yes. Oneness—the sense of oneness. . . . Obviously, if this is to become a natural, spontaneous, and constant condition, death can no longer exist, even in this body. . . . I sense something there which I am still unable to express or grasp mentally. Leaving the body must make a difference, even in the cells' behavior. I need more experience.

If the cellular consciousness is no longer held and imprisoned within the net of a particular body, what happens when that material point, which is in essential continuity with the entire body of the earth, disperses?

63.67 — It's a curious thing . . . the eyesight is totally different from physical vision. You see indifferently thousands of miles away or up close.

72.268 — *(Question:) But what do you see?*
I feel like saying: "Nothing!" I "see" nothing. There is no longer "something seeing," but I *am* numerous things; I *live* numerous things. And then *(laughing)*, it is so, so, *so* much that it's like nothing!

62.117 — Don't you feel something like a sort of pure super-electricity? . . . The minute one feels that, one realizes it's everywhere; we're just unaware of it.

Perhaps "plasma" whose strange properties modern physicists do not fully comprehend?

Such is in short "the other state." We must now try to understand its physiological and functional consequences—"the other way"—as well as the passage between the two states, i.e., what makes the barrier and how to get through it. Clearly, philosophies and religions will not be of any help—exploded.

For centuries, we have been told about "the spiritualists" on one side and "the materialists" on the other, but what matter are they talking about, and what spirit?

What is a fish "spirit" for an amphibian? It is another way of breathing. Pulmonary respiration is what religion and philosophy are all about.

That philosophies and religions are out of the picture is comforting; one less confusion to deal with.

But so is science!

What are the physics, or even astrophysics, of a fish worth to a species from a totally different environment?

All our "laws" in the fishbowl were merely the gauge of our impotence; they were a way of looking, electronically or otherwise, *through* the walls of the bowl. But what happens when the bowl is shattered, when there is no longer any "through"?

Darwin spoke of confessing to a murder.

Mother called that other state "the divine state" or "love," or sometimes "the all-powerful state" or just "that." And even "the supermind."

3

THE NEXT KINGDOM

After all, we might ask, what is the point of becoming a forest or a river if, in our daily life, we continue to stumble and grope for the right action, the right idea, the correct perception, or the genuine intuition? Our human life is besieged by error. What differentiates us from other species is not so much our talent to dissect molecules, invent the radar or probe space, as it is our capacity to make mistakes. Animals do not make mistakes; they know instantly. The entire arsenal of our science is only a huge contrivance to make up for the lack of simple, straightforward knowledge, and to provide us instead with thousands of arms, antennas, and instruments to replace direct action. We are utterly impotent in the midst of a Machine that is supposed to be efficient for us. Should the Machine fail, we would be subanimals.

63.2011 — Something that is not even as harmonious as a tree or a flower, not even as quiet as a stone, nor as strong as an animal—something that is a real downfall. This is really human inferiority.

61.169 — Sri Aurobindo repeatedly said, "Be simple . . . be simple," and as he uttered those words, it was as if a very simple path of light opened up: "Oh, but all you have to do is take one step after another!" Actually, it's as if all the complications stemmed from here *(Mother touches her temples)*. Everything was complicated and difficult to adjust, but then when he said "Be simple," there was like a light coming from the eyes, as if you suddenly emerged into a garden of light. Whenever I see him or hear him, it's like a streaming of golden light, like a sweet-smelling garden—everything, absolutely everything opens up. "Be simple." And I know what he means: do not allow the intrusion of that thinking process, which rationalizes, organizes, orders, and judges, to step in—he doesn't want that. What he calls "simple" is a spontaneous joy in action, expression, movement, and life. In other words, to rediscover in this evolving life the kind of condition he called divine, a condition which was spontaneous and happy.

The New Functioning

We have a very basic thing in common with animals: the cell. Although our amino acids are woven into human proteins rather than mouse proteins, the process is the same. What differs is our mental excrescence, which may be only a temporary excrescence to enable us, consciously and *individually,* to rediscover the power that is subconsciously and collectively concealed in the core of an animal cell. But we have mistaken the means for the end, rather like a crab taking its claws for the supreme organ of knowledge. If there is an evolution and a secret of evolution, if the millions of species strewn upon the face of this good earth since the virus have any meaning at all—and we have to admit there is a progressive meaning in the knowledge of the environment, or successive environments, and in developing a power over the environment, and maybe even in the joy of it, a joy so thoroughly lacking in the case of our species—then, if this meaning, knowledge, power, and joy do not fall from the sky, they must be concealed in the heart of matter's primordial component: the atom and the cell. Only that which is "involved" can evolve, said Sri Aurobindo. The seed, or the atom, already contains its fruit. And the whole significance of our evolutionary journey with its assortment of claws, antennae, vibrating cilia, and cranial protuberances may be to uncover what was *always there*—but momentarily concealed by the main organ we used to explore the outside of the environment. True, we have unlocked the power of the atom, indirectly, through our claws and cyclotrons, but we are still ignorant of the cell's power and the cell's knowledge, because this cannot be approached from the outside; it must be experienced. Our body is the thing we least experience; our head takes up all the space, together with a few more or less happy passions.

And yet, if evolution exists at all, it must take place in matter, in *our* matter.

> **60.65** — At times you feel there's an extraordinary secret to discover, right there, almost at your fingertips; you're about to catch "the thing," to find out. . . . Sometimes, for a second, you see the Secret; there's an opening, then it closes again. Then again, for a second, things are unveiled, and you know a little more. Yesterday, the secret was there, totally clear, wide open, and I saw that Secret. I saw that it is in earthly matter, on earth, that the Supreme becomes perfect. . . .

The "Supreme" . . . what? "Supreme" means perfection of life, perfection of knowledge, perfection of power, perfection of joy—a perfect evolution.

> I saw that Secret, which becomes increasingly perceptible as the Supramental [the other state] becomes more precise, and I saw it in the external, everyday life, the very physical life all spiritualities reject: a sort of precision and exactness down to the atom.

Is it conceivable that this life—so imprecise, so clumsy, so indirect and painful, because it lacks the capacity to implement what it sees—may discover its powerful exactness, its effective knowledge and operative vision within an earthly body? A body aware of each of its millions of atoms, at each thousandth of a second, in New York, Hong Kong, or a corner of this room; in the thousands of beings that live, fly, walk, crawl, or whirl in a cloak of electrons, because that body *is* its own atoms and cells everywhere in the earthly universe, at each second?

This is the "new way" that was developing within Mother's body, and perhaps, through one body, within the entire earth body. We will describe only a few of its most significant stages.

67.23 — The body has become transparent, so to speak, and almost nonexistent. I don't know how to say it. . . . It doesn't oppose vibrations: all vibrations pass through it freely. And the body itself hardly feels its own limits. It's quite a new sensation. It came about rather progressively, but it's quite new, so it's difficult to describe. The body itself no longer feels limited; it feels spread out in everything it does, in everything around it, in circumstances, in people, movements, feelings—completely spread out. It's become very amusing. It's really new. I have to be a little attentive and careful not to bump into things, or when I hold something; the gestures are a bit wobbly. It's very interesting. It must be a transitory condition until the establishment of the true consciousness, which will then function in a completely different way than before, with a precision I can foresee as extraordinary and of a very different order. For instance, in many cases the vision is clearer with eyes closed than with eyes open. But now I see; it's hard sometimes to hold on. It's hard. There are moments of . . . anguish, you know, which would translate into almost unbearable physical pain for an ordinary consciousness. But the result is this that the consciousness of the body itself has really changed; there is no longer anything in there; it's all transparent, like something everything can go through.

71.56 — When the body leaves "that" [the other state], it feels as if it is going to dissolve the next minute, that it's the only thing holding it together. For a long time, one has a feeling that if the

> ego disappears, the being and the form disappear also, but that isn't so! What's difficult is that life's ordinary laws no longer apply. So there are all the old habits, and then there is the new way to learn. It's as if the cells—the organization constituting what we call a human body and holding everything together—had to learn that they can continue to exist without the sense of a separate individuality, although for thousands of years they have been accustomed to that separate existence because of the ego. Without the ego, it all continues . . . according to another law that the body has yet to know, a law that is still incomprehensible to the body. It's not a will, it's—I don't know . . . something—a way of being.

> **67.211** — Now that the cells are becoming conscious, they are really questioning the purpose of all this: "What is the true way like? What is our true function, our purpose, our base? What is the divine way of being? What difference will there be? . . ." And indeed, there is a very faint perception of a way of being that would be luminous, harmonious. That way of being is still quite indefinable, but during this experience there is a constant perception (expressed as a vision) of a multicolored light, of all colors—not layers of colors, but a combination of multicolored dots: a powdering. Now I see that constantly; it's associated with everything, and it seems to be what could be called a "perception of true matter." . . .

There is the old customary matter, as seen through the glass walls of our sealed bowl, and then the other . . . without walls, without the special eyes of a fish or a man—as it is seen by itself, we could say. Though "seen" still implies an external organ: as it experiences itself—matter as it IS—true matter. A perception that would greatly fascinate physicists.

> All possible colors are combined without blending together, and combined as luminous dots. Everything is made of that. It seems to be the true mode of being. I am not yet entirely sure of it, but in any case it's a far more conscious mode of being. And I see it

all the time, with eyes open, with eyes closed—all the time. It gives a peculiar impression at once of subtlety and penetrability, you might say, of suppleness and much less rigidity of form. The first time the body itself felt that in one part or another . . . it was a little lost—the impression of something that escapes you. But if one remains very quiet, this condition is simply replaced by a kind of plasticity or fluidity, which seems to be a new mode of the cells. It could be what will materially replace the physical ego. But the first contact is always very . . . surprising, you know, The moment of the transition from one way to another is always rather difficult. Although it takes place very gradually, there is a moment, a few seconds of . . . suspense, to say the least. And all the habits are undone in the same way. It is the same for all body functions: blood circulation, digestion, breathing—all the body functions. And the transition is not an abrupt replacement of one mode by the other, but a fluid state between the two, and that is difficult. Now I see that for years the body and the whole body-consciousness rushed back to the old way for protection, to escape; but now it has been convinced to stop that, and on the contrary to accept: "Well, if this is dissolution, so be it!" The feeling that all the ordinary stability is disappearing. . . . The great adventure.

It certainly requires a lot of guts.

66.221 — All sorts of minor disorders are breaking out, but the consciousness clearly associates them with the transformation; something knows perfectly well that the disorder has come to help the transition from the ordinary, automatic functioning to the conscious functioning under the direct control and direct influence of the Supreme ["that," the other state]. And when a certain degree of transformation has been achieved on that particular point, another point is taken, then another, and so on. . . . So nothing is done until . . . everything is ready. And it's all a question of changing habits. The whole pattern of automatic, millennial habits has to be changed into a conscious and directly guided action.

67.224 — The difficulty always lies in the transition: if the memory of the other way (the ordinary and universal way of all human beings) comes in, it's suddenly as if—it's quite peculiar—the body were completely helpless and absolutely on the verge of fainting. So it immediately reacts, and the other movement takes over again.

61.26 — It's a very odd thing that came unexpectedly: I no longer knew how to climb stairs! I no longer knew how to do it! Once, something similar happened in the middle of my lunch: I no longer knew how to eat! Naturally, for the outside world, that's what is called "reverting to childhood." Yet, it's truly necessary to let go of everything: all capacity, all understanding, all intelligence, all knowledge, everything—to become perfectly nonexistent. That is very important.

As long as you cling to the capacity and knowledge of the old species, quite obviously you cannot become the other one—it creates an instant wall, the old glass wall of the fishbowl.

69.2112 — This poor body has nothing to say because it knows nothing. It has been clearly shown that everything it thought it had learned in ninety years was worthless, and that everything remains to be learned! So that's how it feels: good-willed, but absolutely ignorant.

70.184 — There are times when the body cannot even stay upright, though the reason is not. . . It no longer obeys the same laws as the laws that make one stand up, so . . .

67.309 — This is the transfer. This morning, for every action, every gesture, every movement, for the behavior of the body and the cells, for the most material consciousness—for everything—the old way was gone. There was nothing but "that," something—how can I put it?—something *even*. There were no more conflicts, no more grating or difficulties; everything was flowing within one and the same rhythm, something so even and that

> feels so smooth, you know, with a FANTASTIC power in the slightest action. That condition was constant and unmixed for about four hours. Now everything, dressing, eating, and so on, are no longer done in the same way. I don't know how to explain it. No more memory, no more habits. Things are not done as a result of the way you have learned to do them; they are done spontaneously, through the consciousness. It means that recall, memory, action are replaced by . . . the new method of consciousness that knows the RIGHT thing to do at the exact moment it does it: "This has to be done." Not: "Oh, I have to go over there." At every minute, you are where you are suppose to be, and when you reach your destination: "Yes, that's it."

When birds leave the Arctic snows for Ceylon's lagoons, they do not "search" for their flying pattern; at every second, they are where they are supposed to be, because . . . the world map flows within them, or they flow within the world's direct geography. We call it "instinct," but that is only our ignorance: the instinct of the world is that it *is* the world, wholly and without separating walls. Then Mother added:

> And so one does understand why the saints and the sages, all those who wanted to feel themselves continuously in that divine atmosphere severed all material bonds: because they were not transformed and so they would fall back into the other way of being. To transform this matter, however, is incomparably superior; it brings on an extraordinary stability, consciousness, and CONCRETENESS: things become the true vision, the true consciousness; it all becomes so concrete, so real [yes, real matter]. Nothing, but nothing else can give that fullness. Escaping, fleeing, dreaming, meditating, soaring into higher realms of consciousness is all very well, but it seems poor in comparison, so poor, so limited!

> **68.45** — The whole solid base that makes a bodily person is gone—just gone, blown away! For example, I have experienced a total loss of memory. Now I am used to it, so the cells remain

quiet, still, and exclusively turned towards the Consciousness, waiting. You see, everything we do, everything we know is based on a kind of semiconscious memory of things—that is gone. None of it is left. And it is replaced by a sort of luminous presence . . . and things happen, one doesn't quite know how. They come effortlessly, JUST what is needed when it is needed. There is none of that baggage we constantly drag around with us—JUST the thing you need.

61.186 — And when the solution must come, it comes: it comes in facts, in actions, in movements.

69.52 — There is no longer that accumulated jumble of so-called knowledge. Everything is spontaneous, natural, and completely unsophisticated; it's very, *very* simple and almost childlike in its simplicity.

70.58 — You see, all the impossibilities, all the "cannot-be's," the "cannot-be-done's"—all that is swept away.

69.263 — The consciousness [of the other state] is constantly working, not as a following of what took place before, but to implement what it perceives AT EACH INSTANT. In the ordinary mental movement, there is always the consequence of what has been done before—not here: the consciousness CONSTANTLY attends to what has to be done; it follows every second—it follows its own movement. That makes everything possible! That's exactly how miracles or dramatic reversals of situations are possible—everything becomes possible!

What if death, illness, physical "impossibilities," "laws"—everything—were but the crystallization of a false memory, that of false matter, that of a certain sealed bowl? A habit going around in circles?

69.2211 — The obstacle is the "concentric" vibration, a sort of concentric vibration, meaning that instead of being part of an infinite eternity, things are viewed in relation to oneself. That's the obstacle. The egocentric stupidity!

62.121 and 64 — It is an extremely delicate functioning, probably because it is not yet accustomed. The slightest movement, the slightest mental vibration upsets everything. . . . For instance, the minute the old way of behaving with one's body (I "want" this, I "want" that, and I "want" . . .) shows its face, everything stops. It takes only one ordinary movement, just one movement of the ordinary functioning—when you slip into that out of habit, everything stops. It's infinitesimal; you don't see those things easily; it's fine, fine, superfine. Then you have to wait until the old mechanism consents to stop. And when you can capture "that" again and stay in it for a few seconds, it's marvelous; then things go astray again, and everything must be started over.

62.2711 — Things are beginning to obey another law. For instance, knowing just to the minute what to do, what to say, what is going to happen—if I pay the least attention or concentrate in order to know, it doesn't work. But if I simply remain like this, in a sort of inner immobility, then the most minute details of life are known at just the right minute: I know what to say, what to reply to a letter; the person who is to come enters. It's an automatic kind of thing you do. In the mental world, you think before doing something; there it's not like that.

70.184 — For example, when I am not supposed to say something, instead of having the thought: "I must not say that," I become unable to speak! And likewise for all kinds of things. The functioning is direct.

66.67 — One always comes back to this: to *be* is the only thing that has power.

Tactile Vision

One may conceive that life could be as spontaneous, "automatic," and harmonious as animal life—that alone would be such an incredible improvement for our species, so encumbered with clocks, doctors, and telephones that it is difficult to imagine. One may conceive that we know at each instant the right thing to do, the right word to say, and everything there is to know in the world, just as the bird "knows" of the lagoon four thousand miles away. But what would be our means of action then, besides flowing within the great rhythm? We differ from other species in our capacity to alter the world, which is something an animal is incapable of doing, probably because it is perfectly attuned to and happy with its routine. Our unhappiness is sometimes our strength. Our uneasy evolutionary detour through the mental fishbowl—where we are cut off and separated from everything, where we have had to contrive and mechanize everything in order to replace a simple missing organ and bring closer that which we drove away in the first place—was probably designed not only to make us individually conscious but, through our very unhappiness, to force us to overcome the "laws" (we haven't; we have merely evaded them, because we do not know their inner workings, the "direct key," as Mother would say) and finally to seize upon the true energy, the lever that will alter the biological routine—something an animal cannot do—and death. The very energy that has formed the galaxies and the cells must surely have the power to change those same cells into a more whole and more stable organism.

The new "organ" of action is very simple, as one might expect. It is neither a mandible nor a new cortical convolution, it is *being.* A type of "being" that has nothing to do with metaphysics, but everything to do with physiology and cellular consciousness. Here again, we will only mention a few stages in the formation of that organ:

64.1010 and 66.263 — For instance, I pick up a page and read it as clearly as I used to before; then the old habit comes back (or simply the thought or the memory) that I usually need a magnifying glass to read—I can't see anymore! Then I FORGET about seeing or not seeing, and I am able to do my work very easily; I no longer pay any attention to whether I see or don't see! And it's the same for everything. . . .

Once again, we are struck by that "memory," which makes you blind, ill, or dying; then the memory vanishes, and the consequence vanishes too! It no longer exists: you see clearly, you no longer have cancer, you do not die. The next species is one that will lose even the memory of death. And Mother added:

It appears incoherent. It must depend on another law that I presently don't know and which governs the physical.

66.93 and 3011 — The perception of people's inner reality is infinitely more precise than before. When I look at a photograph, for example, I no longer see "through" something; I see almost exclusively what the person IS. The "through" is negligible and sometimes noncxistent: I see the photograph suddenly become alive, three-dimensional, with the person's face standing out! It's really peculiar—as if I were being taught to see in a different way.

In other words, eyes and retinas are no longer needed to see, nor is any vision "through" something, as though evolution had manufactured successive organs and successive types of vision adapted to a given environment, then the fishbowl is shattered and one emerges in the "total" environment, with the "total" organ.

65.26 — This vision is rather peculiar. There always seems to be a veil hanging between myself and things [as we will see later, this "veil" is probably the cellular barrier separating us from the other state], then suddenly, for no apparent reason, a certain thing becomes clear, precise, sharp—a minute later, it's gone. Sometimes it's a word shining on a page, sometimes an object. And the quality of the vision is different, too; it's as if the light were inside the object rather than on it. It isn't a reflected light, nor is it bright like a candle, for example; instead of being a projected light, the object has its own light, which does not radiate. That happens more and more often, but without the least sense of logic—or rather, I don't understand its logic. But the vision is extraordinarily precise! And with the full knowledge of what is being seen as it is seen. This morning, for instance, I saw this in the unlit bathroom: a bottle in the cabinet became so bright, so . . . with such inner life! "Well," I thought, "how about that . . ."—the next minute it was gone. Obviously, it is a preparation for a vision by internal light rather than projected light. And it's very . . . oh, it's warm, alive, intense, and so precise! Everything is seen at the same time: not only the color and form, but the nature of the vibration in a liquid—it's extraordinary!

What is this "internal light" in matter, in a liquid? True matter? As it really is, without distorting vision, without any "throughs"?

70.31 and 72.81 — Knowledge is singularly replaced by something that has nothing to do with thought and less and less with vision, something of a higher order which is a new type of perception—you simply know. It is far above thought and above vision, like a perception; there is no more differentiation between organs. And it's a perception which, yes, is global: simultaneously vision, sound, and knowledge. A new type of perception. And you really know. It replaces knowledge. A perception so much truer, but so new that I don't know how to express it.

62.610 — When I look at people, I don't see them as they see themselves: I see the vibration of all the forces that are in them and go through them. That's how I know that my physical sight is not failing but changing in character, because physical details of normal physical sight are false for me! But that doesn't prevent me from seeing physically. For example, if I try to thread a needle while looking at it, it is literally impossible; but if I need to thread a needle, it threads itself! I have nothing to do with it: I hold the thread, I hold the needle, that's all. I think that if that condition perfects itself, it will be possible to do everything in the OTHER WAY, the way that doesn't depend upon external senses; and that would be of course the beginning of a supramental expression. Because it's like an innate knowledge that DOES things.

Perhaps the same innate knowledge that "does" the whole world and every species: an innate knowledge at the heart of every cell and atom? The helium atom "knows" perfectly about its two electrons. And I asked Mother:

(Question:) But wouldn't a "psychic" see in this way?

Not at all! It has nothing to do with the visions I used to have. It isn't a "vision"! I can't even say there's an image—it's a knowledge. I can't even say it is a "knowledge"—it's something that is EVERYTHING simultaneously, that contains its own truth.

63.318 — The sense of "concreteness" is wearing off more and more; it's something farther and farther removed in an unreal past. And that dry, lifeless "concreteness" [our human perception of matter] is replaced by something very simple, very full (in the sense that all the senses participate at the same time), and very INTIMATE with everything. Before, each thing was separate, divided, unconnected with others, and it was very superficial, like the point of a needle. It doesn't feel like that anymore. There is mainly a feeling of intimacy; that is to say, there is no distance, no difference, not "something that sees" and "something that is seen"; and yet, this encompasses the equivalent of vision, sound, touch, taste, and smell—all the perceptions. What prevents this functioning from being perfect are all the old habits. If one were able to let go and not want to "see well" or "hear well," the other, much TRUER perception would be permanent. . . . And always that same feeling of something without conflicts, without shocks, without complications; as if it were no longer possible to bump into anything anymore. It is quite interesting.

72.121 — When it comes, it doesn't come at all as thoughts; it's as if I were BATHED IN IT, and then. . . I don't know, it isn't something I "see"—something outside of myself that I see—it's . . . I AM that, suddenly. And there is no longer any me-you, no longer any . . . I find no words to describe those experiences. For instance, I have lost the capacity of memory, but I feel it's on purpose, that my vision of things would be much less spontaneous and sincere if I were able to remember. That way, it's always like a new revelation, and never in the same manner. That's it: you BECOME the thing—you become it. You don't "see" it; it isn't something you see or understand or know—it's . . . something you are.

66.145 — What takes place here [in our retinal vision of matter, which might be called "false matter"], what we describe is so rough, so lacking in fineness. It is coarse like a poorly carved statue; it's crude, coarse, exaggerated, and distorted by the ego's sense of separation. While *there*—I don't know how to explain it—there all is ONE; it is a single thing assuming all kinds of forms, but not with a center for feeling, another for seeing, and another for understanding. It's not like that—it's a SINGLE substance of extraordinary flexibility that adapts to every movement and to everything that happens, without separation. Afterwards, I am left for hours in a state where I am in this world [ours], and yet I am not. Because I do not feel the way the rest of the world feels. It is a very strange thing.

But that is exactly the vision of the physical continuum!

68.86 — I see now. It's like a oneness, a oneness made of countless—billions at least—brilliant dots. A SINGLE consciousness made of countless brilliant dots conscious of themselves. And it isn't the sum of those dots! It isn't a sum at all—it is one. But an innumerable oneness. The very use of words makes it sound stupid!

64.268 — Everything becomes a LIVING consciousness; each thing emanates its own consciousness and exists because of it. For instance, a second or a minute before the clock chimes, or before someone enters my room or someone even stirs, I know it in the consciousness. And those are not mental things; they are mechanical, yet they are phenomena of consciousness—they are LIVING things, telling you where they are, what their position is. A whole world of small, microscopic events that amount to a new way of living and that seem a product of the consciousness bypassing what we call "knowledge." For example, from time to time I hear people talking about one thing or another, saying, "It will be like this and like that"; it immediately brings on a sort of tactile vision of the thing (how shall I describe this? . . .). It's like vision and touch, yet it is neither; it's both together—it's the thing AS IT IS. A consciousness free of any mental element. And so clear! So minutely precise, like a direct contact with the thing as it is. It is another way of living.

63.411 — It's as if everything were being seen for the first time and from a completely different angle. Absolutely everything: people's characters, circumstances, even the movement of the earth and the stars; everything is completely new and . . . unexpected, in that the whole human mental outlook is completely gone! So things are looking up!

(Question:) But is this the vision of "another world"?

This new vision of things . . . does not mean leaving matter to see the world from a different perspective (that is quite common, of course; all sages and psychics have done that; it's nothing new or marvelous). It's none of that—it is MATTER looking at itself in a whole new way, and that's what is amusing! Matter seeing everything in a totally different way.

The Great Body

We may understand the visual aspect of the new organ, even its tactile character and the direct knowledge it brings, but we cannot help thinking that this is just a somewhat eccentric lady, in her armchair, feeling or seeing "at a distance," through a bizarre kind of tele-vision, and a tactile one at that. But that is because we have not fully grasped the nature of the phenomenon. There is no "distance"; the lady *is* eccentric! An electromagnetic wave is not located in a particular armchair, any more (or any less) than the atoms making up our body are separated from neighboring atoms, except through a temporary, binocular optical illusion—the great separative illusion we live in. All we can say is that a preferential or practical center brings a multiple experience or multiple existence to a certain armchair located in Pondicherry. The center does not dissolve, since it continues attending to its daily chores, chuckling and relating its experiences while remaining in a perfectly physical body, but this center can temporarily be anywhere an action is required—be there in actuality, not only in thought, in vision, or through any kind of "tele"- something, but physiologically and atomically (and in many other ways). Thus we can begin to understand how the supramental being or the next species will operate. The supramental being is primarily an active being, supremely and directly active, contagious, we might say. It is not a super-show that he enjoys in his armchair (though in the present-day conditions, the "show" is anything but enjoyable; it is even painful); it is an immediately transforming action: what is achieved in one's own body is equally achieved in everyone else's body, since one *is* this body, that body, and countless bodies (and not only bodies).

It is best to follow the course of the phenomenon, tentative as it may be, in Mother's own body. Indeed, explanations come afterward; while it happens, it is just odd.

First, this cry:

63.107 — A direct power would be required to change all this!

A power capable of being felt directly, that is, from cell to cell: vibrations of the same quality.

The answer proved to be brutal: a brain hemorrhage—in someone "else's" body.

63.64 — I am conscious of my body, but I don't mean this *(Mother touches her body)*: I am conscious of THE body—it can be anyone's body! I am conscious of those vibrations of disorder that come most often in the form of suggestions. A suggestion of hemorrhage, for example. The body consciousness rejects it. The battle begins to be fought (all this, deep down in the cells, in the material consciousness) between what can be called "the will to hemorrhage" and the reaction of the cells of the body. And it is absolutely like a regular battle, a real fight. But suddenly, the body is seized with a very strong determination and proclaims an order; immediately the effect begins to be felt and, gradually, everything returns to normal. All this happens in the material consciousness. My body has all the physical sensations except the hemorrhage itself, you understand; but it has the sensations, that is, all the sensory effects. All right. Once the battle is over, I look at all that. I see my body (which has been rather shaken, mind you), and I say to myself, "What on earth is the meaning of all this?" A few days later, I receive a letter from someone, and in the letter is the whole story: the attack, the hemorrhage, and suddenly the being is seized with an overpowering determination and hears the words—the very words uttered HERE. The result: this person is cured, saved. I remembered my own event(!). Then I began to realize that my body is everywhere! You see, it is not a matter of just these cells: they are cells in—who knows how many?—perhaps hundreds or thousands of people. . . . It is *the* body! It is so difficult to make people understand that. It is *the* body—this particular one is no more my body than any other bodies. And so it is constantly seized by things like that, all the time; it is besieged all the time, from one side and the other, from all sides.

71.242 — It is off-center, completely off-center. . . .

68.207 — For instance, I don't know how many times a day the following happens: the sudden awareness of a disorder, a discomfort, or a pain somewhere—in some part of the body, but not a part in here *(Mother points to her body)*, rather like in some part of an immense body. Some time later, I learn that a certain person suffered such and such pain—which was felt as being a part of that immense body!

70.281 — I had a rather strange night. The body, the consciousness of the body was that of a dying body, and at the same time with the full knowledge that it was not dying! But it was the consciousness of a dying body, with all the anguish, the suffering—everything—but it was conscious that it [Mother's body] was not actually dying. It lasted a long time; it lasted the whole night. Later, I learned that X had died early that morning. Then I understood. . . .

But that is also how Mother would gradually come in touch with the mechanism of death, and the key to it. For if matter is to be transformed, death is certainly the first thing that must be transformed. That particular key is the key to everything else. Maybe even the key to the sealed glass bowl.

The experience continues:

61.187 — I am inundated with things coming from the outside! And what a mixture! From all sides, all people, and not only from here, but from far, far away on earth, and sometimes far back in time—from the past, things coming from the past to be set straight, to be put in their proper place. It's a constant labor. As though one were perpetually coming down with a new disease and had to find a cure for it.

68.2610 — There are countless experiences, dozens a day, showing that it is that oneness, that identification with other bodies which is responsible for feeling this or that person's misery. . . . It is a FACT. And not as if it happened to another body, but to your own. To the point that now it's difficult to make a distinction. So this body is not bemoaning its own personal misery; EVERYTHING is its misery!

63.289 — This suffering, this general misery is getting almost unbearable, like an acute sort of anguish—which is certainly a necessity in order to find a way out. A way out, meaning a cure, a way of changing it—not an escape. I don't like escapes. That was my main objection to the Buddhists: everything they teach is aimed at helping you escape—it isn't very pretty. But changing, yes.

Changing matter's mortal functioning.

The experience of identification, or oneness, is not restricted to just living human beings, it also embraces the circumstances and "mechanical" events of daily life—in fact, it is all-embracing.

66.179 — There is a new type of activity. I catch myself in the process of doing something: I am talking to people, often whom I don't know, and I am describing a scene to them: if they do a certain thing, it will turn out in such and such a way. They are like scenes from a storybook or a movie. Then, that same day or the next, someone says to me: "I received a message from you, and you told me to write to this person and to tell him such and such a thing!" And I am not doing it mentally at all—I LIVE a scene or describe a scene to someone, and it is received by someone else (and I am not thinking at all about that person!). It happens here, in France, in America, everywhere. It's becoming amusing. Someone writes: "You told me such and such," and it's one of my "scenes"!—one of the scenes I lived. Not lived, but both lived and created. I don't know how to explain it; it's like molding clay. Some stories concern certain countries or certain governments; in those cases, I don't know the outcome—perhaps we'll know it after some time. And during those activities, I know all sorts of things that I don't usually! Sometimes even medical or technical knowledge that I don't have at all! And yet I must know those things, because I say: "You must do it this way or that." It's rather amusing.

64.151 — All this happens in BROAD DAYLIGHT, not when I am asleep. This particular event [one of many] happened just as I was finishing my bath! All of a sudden something comes, takes hold of me, and I live an experience until something is achieved—some action—and once that action is completed, everything vanishes, leaving no trace.

71.177 and 217 — For example, this story about America and China [Kissinger's secret trip to China], and all sorts of similar things, came to me that way. It is very peculiar. A sort of universalization. How can I explain? It's as if I BECAME the circumstances, the people, the words, the . . . The body is more and more conscious, but not at all in a mental way, rather in . . . living through experiences. I don't know how to explain it.

66.1911 — It's not spoken words, nor is it thoughts, yet it's something perfectly concrete coming as if on a screen. And that screen is INSIDE my consciousness—not outside, but inside. And things come in like that. If I were in a superficial state of consciousness, I would ask myself, "Why am I thinking of this?" But I am not actually "thinking," and this isn't a thought, it's . . . a slice of life becoming organized *(Mother makes a molding gesture).* It's very interesting. It ranges from the most minute to the greatest things—cyclones, earthquakes, revolutions—to tiny little things—a little circumstance of life, a donation or a gift that someone sent me—tiny little things seemingly unimportant; everything is stamped with the same value! There is no "great" and "small," "important" and "unimportant." And it goes on all the time like that. It's peculiar. It's almost like . . . a memory in advance.

71.1711 and 70.58 — It's as if the consciousness no longer were in the same position vis-à-vis things, so they appear completely different. The ordinary human consciousness, even the broadest, always occupies the center position, and things exist in relation to that center; in the human consciousness, you stand in one point, and everything exists in relation to that point of consciousness. But now, the point no longer exists! So things exist in themselves. You see, my consciousness is IN things; it isn't something that "receives." I almost feel I am inside all of you, as if I were acting from within you. I don't feel the limits of my body anymore. I don't know how to explain. Yes, it is almost as if it had become fluid. And it isn't, either, like a person expanding her consciousness in order to take others within herself; it's a force, a consciousness that is SPREAD OUT over things. I don't have the feeling of any limits; I have the feeling of something spread out, even physically.

The Supramental Contagion

Thus, we begin to understand the key to supramental action. Or should we say contagion instead of action? Truly a power "from cell to cell."

63.207 — I have a sort of certitude [said Mother at the beginning of her infinitesimal work on the cells, as she was seeking to break through the cellular barrier] that once this microscopic work is over, it will have almost shattering results. Because any action of the power through the mind is soon diluted, weakened, adapted, transformed, and how much of it reaches the lower levels? Whereas when it can act through this matter, it will be obviously overwhelming.

63.107 — Only when this kind of minute work of "local" [cellular] transformation is completed, with a fully conscious and total mastery in the use of the force, without any opposition whatever, only then . . . It's like a chemistry experiment you have learned to perform well: you can repeat it at will every time it is necessary.

61.112 and 254 — *(Question:) How can all the work you do on your own body have an effect on the bodily substance outside you?*

Always in the same way—because the vibration spreads. It's a matter of contagion. Spiritual vibrations are very clearly contagious. Mental vibrations are contagious. Vital vibrations also are contagious (not always in a pleasant way, but it's evident: a man's anger, for instance, spreads very easily). Similarly, the quality of cellular vibrations must be contagious. For example, each time I have been able to overcome something in myself (I mean find the true solution to what is called an "illness" or a bad functioning; the true solution, that is, the vibration that cancels the disorder and sets things straight), I always have found it very easy to cure those people who had the same thing by sending out that vibration. It is so because the entire substance is ONE. Indeed, everything is one; that's what we keep forgetting! We live with a constant sense of separation, but that's a complete falsehood! Because we rely on what our eyes see—that's *really* falsehood. It's like a picture plastered over something. But it isn't true. Even with the most material form of matter, even with a stone, the moment you change your consciousness, that whole sense of separation and division totally disappears. There are only different . . . what shall I say? . . . modes of concentration, or modes of vibration, WITHIN THE SAME THING.

64.73 — X was in an intense state of emotion and, at a certain point, our eyes met. The emotion coming from him entered me with such violence that I almost broke into tears, believe it or not! And it's always here, in the lower abdomen, that that identification with the world takes place. I immediately stopped X's vibrations (it took me a few minutes), and everything returned to normal. But I understood that that kind of contagion is kept active as a means of action—it isn't too pleasant for the body! When I put things in order here *(pointing to the abdomen)*, it also puts things in order there.

63.1112 — When the experience [of the other state] comes, it is rather widespread: "that" flows in the blood, quivers in the nerves, vibrates in the cells and everywhere, and not just in the cells of this body; I feel that the blood, cells, and nerves of many bodies participate. Naturally, the central consciousness of the person may not always be aware of it, the person may not be aware of it (he experiences an extraordinary sensation but does not know what it is); whereas the cells know what it is, but they cannot express it. There are DEGREES of consciousness, you know, and this *(Mother's body)* seems simply to be a more conscious center of consciousness, but other than that . . .

Then the experience becomes more and more precise, universal.

68.186 — Curiously, while following a certain movement, I . . . take off [into the "undulation"]. That can happen at any time at all. While I am eating, in the middle of my lunch, something comes, and I follow the movement, with my spoon in midair, and then I realize that everybody is waiting!

(Question:) Yes, for months now I have noticed that: like a remoteness on your part.

Not at all! I am INSIDE, much more so than before. Not "inside" here *(Mother points to her body)*; inside everything. Whenever I go off like that, it's always as if I were . . . molding vibrations. Later, I learn that something happened to somebody, something went awry; so one tries to straighten it, to bring back the light, the right vibration.

64.269 — I am talking here about the body's cells, but the same thing applies to external events, including world events. It's even remarkable in the case of earthquakes, volcanic eruptions, etc. It would seem that the entire earth is like the body.

62.237 — It's becoming more and more a general yoga—the whole earth. And it goes on day and night, when I walk and when I talk and when I eat: like kneading dough and making it rise.

Finally, the experience became perfectly clear and, one morning, Mother exclaimed:

61.2312 — I had a perception of the power, the power that comes from supreme love [the other state]. Fantastic! It made me

> understand that I am being put into this condition to acquire that power, which results from an identification with all material things. And I looked at that power from a practical standpoint, to organize matter, not accidentally or spasmodically as mediums do, but for a real ORGANIZATION OF MATTER. Then one begins to understand: why, with "that," everything can be put into its place!—provided you are sufficiently universal. It's incredible! It has the power to change everything, and in such a way! One simply IS "that," a single vibration of "that." Indeed, one IS "that," and hence one DOES "that." Why, that's the key!
>
> **58.262** — A direct key that requires no elaborate science to express itself.

We could say that our whole mental, or even animal, kingdom is the "indirect" kingdom, the kingdom of mechanisms: from the shrewmouse using its teeth to chew a vine to our physicists smashing atoms in their cyclotrons. Countless and increasingly intricate mechanisms, from vibrating cilia, wings, fins, to jet propulsion and telexes. A gigantic artifice. It is as if evolution, that is, a certain energy (and energy goes with consciousness, even the "consciousness" of the hydrogen nucleus snatching its single electron) had garbed itself with more and more adapted and ingenious mechanisms or organs in order, finally, to reach that evolutionary turning point when the mechanism becomes conscious of its own engine, stops dividing itself into countless bodies, becomes one again with the totality of its substance from galaxies to atoms, and starts operating directly on its substance, its nuclei, cells, and all universal matter. After the mineral, vegetable, and animal kingdoms comes a future, "direct" kingdom: a reorganization of matter through matter's own power, and through the consciousness contained in each atom and cell. But one must reach *that* point—the atomic and cellular level—rather than soaring to nirvanic or celestial heights; one must break through the barrier that separates us from the future "total" environment, from our next global species, just as one day the mineral form broke through the barrier of its inertia. The beginning of evolution meets the end: power retrieves its power, and unconsciousness its buried consciousness.

"Salvation is physical," Mother said.

68.1112 — The body is a very, very simple thing, very childlike, and it has that experience in such a compelling way, you know; it does not need to "seek" anything—it's right THERE. And so it wonders why men never knew of this from the start: "Why, but why did they go after religions, gods, and all those . . . sorts of things?" While it is so simple! So simple! It's so obvious for the body.

64.3010 — All the mental constructions that men have tried to live by and realize on earth come to me from all sides: all the great schools of thought, the great ideas, the great realizations, the great . . . and then, lower down on the scale, the religions—oh, how utterly childish all that seems! And a sort of certitude in the depth of matter that the solution lies THERE. Oh, what a fuss! How you have tried in vain! Go down into yourselves, deep enough inside, remain there quietly enough, and "that" will be. And you cannot understand; there's only to BE it.

61.182 — *(Question:) But why is it necessary to go inside? Why not act on matter from above?*

Act from above? I have acted from above for more than thirty years! It doesn't change anything! It doesn't transform. Transforming means transforming. To transform, you must go down into the body, and that's the terrible part. . . . Otherwise nothing will ever change; things will remain the same. You see, it's even fairly easy to pose as a superman! But that remains ethereal, it isn't the real thing, not the next stage of terrestrial evolution.

62.245 — These two positions, the spiritual position and the materialistic position—each of which considers itself to be unique

> and exclusive, and thus denies the other's validity in the name of truth—are inadequate, not only because they don't admit each other, but because to admit and to integrate both is NOT ENOUGH for solving the problem. There is something else—a third position that does not derive from these two but is something to be found that will probably open the doors to total knowledge. That "something" is what we are looking for—perhaps not only looking for, but MAKING.

A new *physiological* position in matter. No longer a philosophical position with its so-called materialisms and spiritualisms, which are but the two sides of the same false vision of matter, but a position of the body, in the body, that will change all the laws of the old "frame of reference."

A new mode of life in matter that will reorganize matter through its own power and will ultimately change death; for death was only the reverse side of this life, just as the other side of the fishbowl was not the death of the fish but the beginning of a new life form in matter.

And, peering into the future, we begin to catch a glimpse of the supramental being's mode of action, how he will operate in matter.

> **58.192 and 32** — Whenever some change is needed, it is not made through artificial and external means but through an inner operation, THROUGH AN OPERATION OF CONSCIOUSNESS that imparts form and appearance to the substance. Life creates its own forms. . . . What is absurd, here, are the many artificial means you need to use: any fool has more power if he has more means to acquire the appropriate devices. In the supramental world, on the contrary, the more conscious you are and attuned to the truth of things, the more authority you have over the substance. The authority is true authority. If you want clothes, you have to have the power, a real power, to make them. If you don't have that power, well, you go naked. There is no device that can substitute for the lack of power. Here, authority is not once

in a million the reflection of something true. Everything is incredibly stupid!

Supramental consciousness imparts form to matter; it molds matter by releasing the appropriate vibration, just as today we mold thoughts through words.

But how do we *get* there? What is the procedure?

4

THE DESCENT INTO THE BODY

Our question is really about death. As long as the physical reality of the coffin or the funeral pyre is not changed, nothing will be changed, and we will perpetuate the "law" that has ruled life since the protozoa, even if we temporarily take off into the "undulation." "It's almost as if it were *the* question given me to resolve," said Mother. Mother is first and foremost the struggle against death, because Sri Aurobindo had died in 1950. Like Orpheus and Eurydice. For twenty-three years, she was to battle with "the question," like a lioness. But in fact, one *cannot* go into the undulation and ubiquitous life unless something has changed in the system of death, because what creates the barrier also creates death. And what does create that barrier? What is the cellular mechanism of death? Scientists observe the features of the phenomenon and declare: the combination of this plus this plus that causes death. But *why* is there "this" in the first place? They have no idea. Above all, the next kingdom will have a different position vis-à-vis death—it will be outside it. If the mode of life is to change, the mode of death also must change, or else it will just be the same old story, with a few ubiquitous and heavenly illusions in between.

Where is that nest of death to be found?

The only way to know is to descend into the body.

That's what Mother meant in 1959 by "the path of descent." That may even be what the "descent into hell" is.

A Mortal Habit

But how does one go about descending into a body? Our body is perfectly natural to us. We walk on its two feet and simply keep it fed, rationalizing the whole thing with a few convenient layers of philosophy or verbal acrobatics. And off it goes. If we are to begin to understand anything about the body, it must, first of all, stop being "natural." As long as the guinea pig behaves normally in its cage, it will only go on producing guinea pigs that will produce more guinea pigs. One can try changing one's diet, one's sleeping patterns, or even the quality of the air one breathes; one can even stop one's heartbeat. Hatha Yogis have extensively experimented with all these techniques. And then what? We are not after the acrobatics of the old species, or even after an "improvement" of the old species. We are after something else. We can tinker with all the body mechanisms, but that will not yield anything really new, because these mechanisms barely touch the surface—and this is why neither biologists nor Hatha Yogis have found the key, nor even really understood the problem. As we said earlier, from the time of the shrewmouse we have done nothing but juggle with one mechanism or another. We must find another, deeper key within the body.

But how do we get there?

The experience is simple really—simple to describe in any case—but it is meaningless unless it is *lived,* because it is not in the pages of a book that one changes the functioning of the body. We are not attempting another theory, but a new *creation.*

Combining biology and yoga, we might imagine that that "descent" is going to reveal a buzzing network of nerves and veins, that we will hear palpitations and vibrate with the nucleoli and dendrites—in short, experience our body at a microscopic level through some kind of electronic or yogic magnifying glass that will eventually reveal the "key." But the key lies in none of these good things—our body is an excellent body, as good as that of a snake or a kingfisher, with only minor mechanical differences. So what prevents this excellent thing from being so excellent? What causes it to turn into a human being's body rather than a beetle's—both mortal, it should be noted? Biologists, who always answer in terms of operating processes because it is the only thing they can grasp, will reply that this body is a human one rather than something else because certain amino acids—the same from the primitive virus all the way to Einstein—are assembled or "woven" in a certain way, in a certain order, yielding human proteins rather than something else. There is no escaping it; it is implacable and scientific from the first hydrogen cloud onward: they will go on being woven that way, or slightly differently, forever. "Materialism is the gospel of death," Mother said strikingly (and since spiritualism is the gospel of heaven, we really have to find something else that will sit a little better, and live a little better, between these two extremes). But *what* weaves them in a particular way? What is the prime mechanism, the underlying force that causes them—or forces them—to be woven at all, without great difference between Homo sapiens and a lizard? It is not the difference between a lizard and a human being that is of interest, but the fact of the weaving itself. What law does it obey? That is what the scientists do not know. But Mother does. This is the "key" that interests us. If we had that secret, we would not start weaving the proteins in a different order so as to give birth to some questionable new species, but we would possess the key to life itself: why it becomes a fish or a human being, or rather why it gets stuck in a particular typal habit, and why it dies. From one species to another, there is only a different habit in weaving the same materials. What controls that habit of matter? This is the real question. If we find that answer, then we may find what causes us to die, and perhaps we will lose the habit of dying.

A certain habit.

The Mental Layers

Thus, that descent into the body is not done through any yoga technique, but in the simplest possible way: one goes into what is right there. And one does not encounter any network of veins and dendrites but something else altogether, which indeed looks like a strange Amazonian jungle. In order to perceive or experience the cells, one must first break through, one after the other, the opaque, buzzing layers that are covering them.

The first of those layers is our *intellectual layer,* the one we ordinarily live in. That is the top of the mental fishbowl. All the ideas, philosophies, religions, and so on obviously have nothing to do with the body. We do not even notice that layer; it is like the very air we breathe, yet it is a huge seething mass. All that must be silenced. Just as a liquid first has to settle in order to be able to see through it. The first step is therefore to silence the mind.

When that layer is somewhat clarified, a second layer comes into view with greater sharpness, as it is no longer embellished by the higher noise of great ideas and lofty philosophical concepts. This is the layer of the *emotional mind.* This layer is noticeably stickier. Yet emotions, however lofty they may be, have nothing to do with the body. Hence, the second step is to quiet the emotional mind. This is a more complex task, somewhat like guerilla warfare in the desert. When that layer becomes a little clear and calm, a third layer appears that, up to now, had been totally blended with the first two: the *sensory mind,* which controls our reactions. Now, this is really the thick of the jungle teeming with all kinds of little snakes and swamps. All the sensations of fatigue and sleep, fear, pain and pleasure, like and dislike, attraction and repulsion, contraction and release—a teeming mass. We have not quite reached the body, but we are getting close. And one does realize to what extent all that is the result of habit, of the environment, of education—a hodgepodge that has nothing to do with the body but only clings to it, as it were. The third step, therefore, is transparency of the sensory mind, or perfect neutrality. The least contraction or rejection is like putting up an instant wall: the progress stops; one remains stuck in the middle of the Amazon. The body must be freed from all that network of actions and reactions. Now, the body begins to feel slightly wobbly, as if it has lost its moorings and is unsure of its mass. In reality, it is delightfully unburdened and light; it begins to feel like "the body." Finally, we reach the barrier: the fourth layer, the *physical mind.*

But we don't know that it *is* the barrier, or where we are or what we are doing in that strange jungle—it's only *afterward,* after we have crossed it, that we realize it was the barrier and what it meant. At the time—and "the time" lasted years for Mother—it's only an endless, microscopic, sticky, seething magma, which we don't know if it leads to the "other side" or to the body's disintegration, or even if there exists another side to that microscopic hell which adheres so tightly to the body that it seems as if any attempt to pull it off would mean pulling apart the body itself. When Francesco de Orellana went down the Amazon River for the first time, from the Andes, it wasn't called the Amazon then, but "a place" with creeping vines and alligators. He did not know whether it would lead him to the Atlantic Ocean or to death, and he had no idea of what he was crossing. It is easy to draw maps *afterward.*

We will give only a few stages or landmarks of that crossing through the layers to the barrier of the physical mind.

> **65.247** — When you pay close attention, you realize that what takes the longest is to become conscious of what needs to be changed, to have a conscious contact enabling the change.

How long did it take the great primates to realize that what counts is not swinging from branches but sitting thoughtfully in a corner of the brush and gazing at . . . nothing?

> **66.303** — If you want to experience the body, you must live in the body! That's why the ancient sages and saints didn't know what to do with their bodies—they left them and went on meditating, so their bodies didn't participate at all.

> **63.108** — It's a monumental battle against habits going back thousands of years.

> **59.195** — When it comes to the body and you want to make it take just one step forward—oh, not even a whole step, only a tiny one!—everything starts grating; it's like stepping on an anthill.

> **56.276** — The moment you want to progress, you immediately encounter the resistance of everything that refuses to progress within you and around you.

> **58.256** — There is such a long way to go between the usual state of the body—this almost total unconsciousness we are used to because "that's how it is"—and the perfect awakening of the consciousness, the response of all the cells, of all the organs and body functions. Between the two, there seem to be centuries of work.

53.1410—Death isn't inevitable. It is an accident that has always happened until now (or seems to have always happened), and we have got it into our heads to conquer and overcome this accident. But that entails such a terrible, stupendous battle against all the laws of nature, all the collective suggestions, all the earthly habits, that unless you are a first-rate warrior ready to go through anything, you had better not begin the battle. Absolute fearlessness is required, because at every step, at every second, you must wage a war against everything that is established. So it isn't exactly easy. Even individually, you have to battle with yourself, because if you want your physical consciousness to reach a state allowing physical immortality, you must be so free from everything the present physical consciousness stands for that it's a battle every second. Every feeling, every sensation, every repulsion, everything that makes up the fabric of our physical life must be overcome, transformed, and freed of all its habits. That means fighting every single second against thousands and millions of adversaries.

63.3010—The body is starting to understand one thing: EVERYTHING that happens is for progress. Everything that happens works toward the attainment of the true state, the state that is expected of the cells so the realization is accomplished. Even blows, even pains or apparent disorders are all for a purpose. And it's only when the body, stupidly, takes it the wrong way that it gets worse.

60.281—The difficulties come from very small things that seem absolutely commonplace—yet they block the way. They come without reason—a trifle, a word, an illness afflicting someone close to me; suddenly there is a contraction, and all the work has to be done over again as if nothing had been done. One might think that the form of the body is a point of concentration, and that without that concentration, or rigidity, physical life would be impossible. But that isn't true! The body is really a wonderful instrument; it is capable of widening, of becoming vast. Then everything, the slightest gestures, the slightest work is done in marvelous harmony, with a remarkable plasticity. But suddenly, over some trivial detail, a gust of air, a nothing, it forgets; it shrinks back onto itself out of fear of disappearing, of no longer being. And everything has to be started again from the beginning.

61.157 — To be a saint or a sage is not that difficult, after all, but supramental transformation is something else! No one has ever taken that path; Sri Aurobindo was the first, and he left before telling us what he was doing. I am literally cutting a trail through a virgin forest—worse than a virgin forest. And so I have a feeling of knowing nothing at all. From the purely material standpoint, that is, chemically, biologically, medically, or therapeutically speaking, I don't think many people know (maybe there are some; in any case I certainly do not). From the yogic standpoint, it is very easy: you know what you have to do, and you do it easily, there's nothing to it. But that transformation of matter is truly something else! What has to be done? How is it to be done? What is the way? Is there a way? A procedure? Probably not. I am given the awareness of how enormous this task is one drop at a time . . . so I won't be crushed. It has come to the point where all spiritual life, all those people and races that have tried since the beginning of the earth—all that seems like nothing, like child's play in comparison. And it's a work without glory: you have no results, no experiences filling you with ecstasy or joy—none of that; it's a dreadful labor. It's truly like walking in nothing, with nothing, in a desert strewn with every conceivable trap and obstacle. You are blindfolded and you know nothing.

If the eyes of the body are to open, the eyes of the mind must close.

60.165 — Up above, everything is fine, but down below it's a swarming mass. In fact, this is a battle against very tiny things: certain habits of being, certain ways of feeling, of reacting . . .

69.2712 — When it comes to material things, intelligent people instinctively feel it's a known and solid ground, based on established experience—and that's just where one is vulnerable. That is precisely what the body is being taught: the inanity of our present way of seeing and understanding things, based on right and wrong, good and evil, light and dark, all these contradictions; our entire appreciation and conception of material life is based on them. Even the physical part of ourselves—which believed it knew how to live, what to do, and how to do it—must also understand that that isn't real knowledge, that isn't the true way of handling external things. For instance, the consciousness that is at work is constantly "teasing" the body as if to say: "See, you feel this way; well, what is it based on? You think you know, but do you *really* know what's behind your feeling? . . . and so on and so forth for every little thing of daily life. There is a constant living demonstration that whenever you do things from a sense of acquired wisdom, acquired knowledge, or past experiences, how . . . false it really is, and that there is really SOMETHING ELSE behind.

58.105 —You are beaten and battered until you understand, until you are in that state in which all bodies are your body.

The moment the slightest reaction of "me" intervenes in the body, it raises an immediate wall—the old story of the one-celled organism and its protective membrane.

60.1211 — A more and more total and integral consent, a more and more accepting surrender . . . One really feels that the only way is to be just like a child. If you start thinking, "Oh, how I wish I'd be this way! Oh, I ought to be that way," you're just wasting your time.

How could one possibly know how to be the next species?

60.1712 — Sometimes you feel, "That's it! I've got it." But back it falls—back to toiling. Sometimes you feel you're falling into a hole, a real hole, and how do you get out of it? It goes on like that weeks on end. But what is most striking is that the sense of the "important" and "unimportant" is disappearing. You're left like that, with . . . nothing. There is no scale of importance! That's entirely our mental stupidity, because either nothing is important, or else EVERYTHING is equally important. The speck of dust you wipe off the table, or ecstatic contemplation—it's all the same.

Indeed, what is "important" for the next species? We will know it only when we get there. The coccyx vertebrae are an unimportant vestige from an organ that was quite important to monkeys.

62.610 — It is easy to understand: if it were a matter of ending one thing and beginning another, it could be done fairly readily. But keeping a body alive, and making sure it continues to function, while at the same time pursuing a new functioning and a transformation—that makes for a very difficult combination to realize, especially if you take the heart, for example. The heart is to be replaced by the center of Power, a tremendous dynamic power! At what MOMENT do you stop the blood circulation and throw in the Force? That is not easy. In ordinary life, you think things out, then you do them. Here it's just the opposite! You must first *do* things, then you understand them, but much later. You must first do without thinking. If you think, you don't achieve anything; you are simply back in the old rut.

62.3010 — It is new, which means that you don't know whether you are progressing or not! You don't know where you're going, or what path you're on. There are all sorts of things happening, but are they part of the path or aren't they? I have no idea. Only at the other end will we know.

63.226 — This transitional period is truly not satisfying, in the sense that you no longer have the strength and capacity you used to have, yet you have none of the power and capacities you expect; you're midway, neither here nor there, with absolutely amazing things happening. At times, some things leave you totally wide-eyed: "So that's what it's like?!" But then, also, such tiring limitations, so tiring!

71.2912 — For me, the shortest way was—how can I say this?—a growing sense of my own inanity, my nonexistence. The realization of my helplessness, my unknowingness, the surrender of every ambition. But there is no place for fear; if you're afraid, it becomes dreadful. Fortunately, my body is not afraid.

65.1010 — It is the mass of all the things labeled unimportant—that's what prevents the physical transformation. And because they are such tiny things considered negligible, they are the worst obstacles. I am speaking about people with an enlightened consciousness, who live in the truth, who have an aspiration and wonder why that intensity of aspiration produces such meager results—now I know. The meager results are due to the lack of importance given to those tiny things that are part of the subconscious process, so that you may have achieved freedom in your thoughts, in your feelings, and even in your impulses, but you are physically enslaved. All that must be undone, undone, undone. It is nothing but mechanical habits. But it clings; it's really sticky, oh! . . .

67.267 — We could call our world a world of bad habits.

67.28 and 19 — A slow, invisible, almost imperceptible underground labor. A sordid battlefield.

65.257 — What I call sincerity is to be able to catch oneself at every minute being part of the old stupidity.

65.121 — In the old days, you were told: "Don't pay any attention. Just let that seething mass drift about." But we don't have the right to do that! It would be contrary to our work. I had achieved almost total freedom with respect to my body, you know, to the point that I was able to be totally immune to physical pain; and now I am not even allowed to leave my body, if you can imagine! Even when I am in pain or when things are rather difficult, and I say to myself, "Oh! to go off into my blissful state"—it isn't allowed. I am tied down here. It is here, DOWN HERE, that the realization must be done.

60.2611 — Things are not at all the same as in ordinary life, yet for three or four, sometimes ten minutes, I am horribly sick, with all the signs that it's over. But it's intended for me to have the experience, so I can find the strength. It's only in "those moments," when logically, according to ordinary physical logic, it's the end, that you grasp the key. It's just a question of going through it all unflinchingly. How many more such "moments" will be necessary? I have no idea. I am hewing out the way.

There has to be the last convulsions of a reptile at some point in history to begin finding the key to the bird.

69.35 — Death, food, and money are the three great "dominants" of human life according to that new consciousness. Human life revolves around these three things: eating, making money, and dying. Yet, for that consciousness, the three are . . . temporary inventions, the result of a transitory condition, something neither very permanent nor very deep. And so it is teaching the body to be otherwise.

61.125 — Even those moments in life when you have sudden glimpses of an immortal consciousness, a contact with a truth—all those experiences are well and good, but they are not IT. What is the true SENSE of life? What is life really about? What is behind it? Why has the Lord done that? What is His goal? He evidently has a secret and He is keeping it. Well, I want His secret. Why are things the way they are? They certainly aren't this way just to be this way; they're meant to be something else. That's the something else I want.

62.2311 — Every step forward forces you to take a step not backward, but into darkness, and in physical terms it's terrible. It is like touching a bottom of unconsciousness and of . . . inert matter.

63.218 — I don't know whether this is the last battle, but it has reached very deep. It's like the primary substance used by Life, with a sort of incapacity to feel or experience any reason to this

Life. I have the feeling it is quite close to the bottom. At one point, there was such dreadful anguish, because of the utter nothingness—a nothingness you couldn't get out of. There was no escape from that nothingness, because it *was* nothing. For an instant, the tension was so great that . . . you wonder: "Am I going to explode?" And this is the base, the foundation of all materialism.

Then, suddenly, the barrier appeared:

61.157 — Every possible difficulty in the body's subconscious rose up en masse—as it had to happen, and as it must have happened to Sri Aurobindo. Now I understand! Well, it isn't a joke, you know! I had wondered why all those things were furiously after him. Now I know, because the exact same fury is after me. It is not exactly the consciousness of the body, but the bodily substance as organized by the mind, one could say: the very first mental movement in Life. What caused the transition from animal to man, you know, the first mentalization of matter. Well, something there is protesting, and the protesting produces physical disorders, naturally.

We are at the threshold of human life, facing "something" that does not exist in animals and that creates all the complications of human existence—its nonknowledge, its pain, its separateness, its illnesses—all the "misfortunes," which, in the end, are the true lever for getting out of them, because they force us to go to the root of things to find the key. This is the physical mind, the first "mentalization" of matter. This is the barrier. And at the same time, it is the opening on an even more radical discovery, a deeper layer yet: that of the cellular mind, which holds not only the power to undo our habits of unhappiness, but to undo every species' typal habits and, ultimately, the old habit of dying.

5

THE PHYSICAL MIND

This physical mind is an extraordinary discovery. And yet, it is right in front of our eyes, droning in our ears and controlling our slightest gestures. Only we do not notice it; or if we do, we soon discard it as insignificant or we drown it in the bustle of our lofty thoughts, lofty feelings and the rest of our lofty attributes, which all crumble precisely because that microscopic creature was never taken into account. The greatest discovery is finding what blocks the way. If each species knew what blocks the way to the advent of the next species, it would soon overturn all its values and find the passage. For that, however, a certain discomfort or a beginning of suffocation within one's species is necessary; such is our privilege among all the animals that happily go around in circles in their own various sealed bowls. If some fish had not begun to suffocate in their drying ponds, they would never have invented pulmonary respiration and changed their fins into paws to produce amphibians. That physical mind is precisely what is suffocating us—insidiously, multifariously, and very implacably. It is our cage, the very wall of our human fishbowl. We do not need extraordinary mutations to leave our fishbowl; all we need is enough suffocation to find the way out. And perhaps our species is indeed reaching its time of suffocation.

The "top" part of that physical mind, at least, is familiar to us. It is the one we hear repeating microscopic material thoughts ad nauseam, like an old crone talking to herself. If we didn't check it, it would go on repeating for hours, relentlessly: "You haven't closed the door; go and make sure . . ." like a broken record, and you know perfectly well you've closed that door! It repeats everything—the slightest gesture, the least word, the tiniest tripping on a stair—and it remembers everything twenty years later, in exact detail. An implacable memory. The size of a pinhead, it seizes upon any subject matter, digs its furrow, and then repeats forever and ever. In our every nerve, from head to toe, and even in our cells, we are scored by that unruly mechanism. In fact, we are literally made up and covered by that physical mind. It is the great fixative, without which we might forget that we are humans, forever yoked to this mode of matter, and to death. But its role is precisely that: to harness us to matter.

Its second "feature," which we can perceive somewhat in its higher and more visible parts, is fear. It is afraid of just everything: "Careful! You forgot to take your scarf; you'll catch a cold. Careful! If you go too fast, you'll break a leg. Careful! You can't do that, or you'll tire your heart." You cant, you can't—it's a mental creature teeming with can'ts. Even if you could, it would prevent you, and that is why you can't in the end. In short, it is the keeper of the fishbowl's boundaries. It is the prison warden. "The doctor said . . . " and the professor said and Webster's dictionary said, and so did the sheriff, the minister, and the biologist—isn't that enough? Everyone says so—isn't that doubly enough? It is the world's greatest logician. A many-faceted, microscopic, implacable logic. The physical mind is the greatest sheriff of all species: "Be realistic. You can't possibly leave the fishbowl! There isn't any material water on the other side, only death and fish Spirit. And in any case, it doesn't really exist; one can't swim in it or touch it or see it. So isn't that enough?" But its "logic" leads straight to the desired nest—death. Everything leads there. This has nothing to do with the survival of the species; it has to do with the survival of death. Just follow its microscopic whispers. Whenever there is the slightest scratch: "Oh, isn't this poisonous?"; or the least sneeze in Moscow: "Oh, will there be war?" It foresees every possible disaster, every possible disease, every accident. And above all, death; it has foreseen death from the start. "This is a DISEASE; there's no way out of it! You MUST take this many pills, you must do this, you mustn't do that." We are bound from head to foot, invisibly, insidiously, and inexorably. It is a fear-of-everything embedded in matter, like a memory, or a regret, of the happy, peaceful inertia of the stone; life is a calamity to that inertia, a threat, a perpetual danger. Death is peace at last. So it spins and secretes its little death at every second, until it gets what it wants: "I told you so." And what would the ecclesiastic hierarchy do without death? And the biologist, the philosopher, and the rest of the pack? They all live on death. So it is the keeper of the law of death. Any law means death. It is the gospel of death from A to Z. The ultimate example of that creature's doings can be found in a person afflicted with the uncontrollable trembling of Parkinson's disease who tries desperately to take a step, trips, and tries again—"You can't do it; you see, you can't walk"—until the person is "fixated" once and for all. Its job is to fixate. Thus, we realize the incredible hypnotic power of that physical mind; it really takes all our mental din for us not to notice the omnipotence of the infinitesimal whisperer. In fact, this is the domain where all professional "healers" and "hypnotists" operate when they are able to prevent feeling a pain that would be otherwise excruciating, or when they make a person do "impossible" things that contradict all the "cannots": they abolish, for a moment, the physical mind. Doctors, also, act on the physical mind, sometimes to cure an illness, but most often to "fixate" it forever. In our higher consciousness, we laugh at that harping and ever-fearful caricature, and we throw it out the window—but it remains underneath, spinning its little deaths, diseases, and accidents, which *will* culminate in the big, quiet, and final death. In the end, it always catches up with us. Something in living matter yearns for the peace of the mineral state. There it is an implacable memory going back to the beginning of time, perhaps to that original condition of matter in which the supreme power is buried in what appears as utter powerlessness and immobility in the supreme movement of the atoms. If the death of the species is an obstacle, it must be the key to something else. Wherever there is a wall, there is also the other side of the wall. The only real obstacle is in not seeing the wall.

We will give a brief account of Mother's crossing through that last layer which wraps us tightly and hermetically and "locks" us in our human, mortal way. That is what Mother called the "horrible thing." We are actually wrapped in a fourfold superimposed web: the first, relatively loose-meshed web of the intellectual mind; the second, with a somewhat tighter and stickier mesh, of the emotional mind; the close-meshed web of the sensory mind; and finally the microscopic mesh of the physical mind. Underneath all that is the body, i.e., an unknown whose reality totally eludes us, because everything supposedly arising from the "body" is distorted, falsified—manufactured really—by the four successive webs. So what is actually underneath? Biologists may talk of enzymes and DNA molecules, but it is like talking of a man forever locked up in a dungeon. Let him, first, get out of his dungeon and feel the sun, and we'll soon find out whether the little molecules still behave the same way, and if all the "laws" of the biologists were not merely the laws of the dungeon.

54.103 — They would rather die and keep their habits than live in an immortal way and give them up.

57.155 — I challenge you to change your body if your mind is not changed. Just try and let's see what happens! You can't move a finger, say a word, take a step without the mind intervening—so with what instrument do you expect to change your body if your mind isn't changed beforehand?

58.105 — One of the most serious obstacles is the sense of legitimacy that the outer, ignorant, and deceitful ordinary consciousness gives to all the so-called physical laws—causes, effects, and consequences—and to all the physical and material scientific discoveries. All that is indisputable truth to our consciousness, and it is so automatic that we're not even aware of it. Where such things as anger, desire, etc., are involved, we see that they're wrong and must disappear, but where material laws are involved—those of the body, for instance, with its needs, its health, its nutrition, and all those things—they have such a solidly concrete reality for us [exactly: the dungeon], they are so solid, so established, that they appear absolutely irrefutable.

61.173 — Everyone is locked inside a small formation created by the most ordinary mind that molds everyday life, as if in some narrow prison.

67.2110 — And then there are all those old things arising from human atavism: being reasonable, being careful, being smart . . . taking precautions, providing for the unforeseen—oh, the whole tissue that makes up the ordinary human equilibrium. It is so sordid! The whole mentalization of the cells . . .

The cells are "mentalized," i.e., hypnotized, and perhaps even terrorized by the prison warden.

The whole mentalization of the cells is like that, full of it; and not just because of one's own way of being, one's own experience, but also that of one's parents, grandparents, friends, and so on.

> **68.2610** — It really is a kind of hell. Only that Possibility [the state outside the dungeon] prevents it from being hell, otherwise . . . The impression, you see, is that all the layers of the being have been sort of whipped together (you know, like mayonnaise); so naturally, with all the layers well blended together in the greatest confusion, the "horrible thing" is bearable because of everything else that's mixed with it! But if you separate it from the rest . . . It is quite clear that if it weren't unbearable, it would never change.

Mother lived in that last "pure" layer separated from the rest, at the threshold of the body, searching for the passage.

> **62.62** — It is such a dull, stupefied consciousness. It gives the impression of something immovable, unchangeable, and totally unresponsive; the impression that you could wait thousands and millions of years, and nothing would budge. It takes cataclysms to get it to budge a little, it's really extraordinary! Not only that, but the wisp of imagination it does have is always disaster-oriented. If it foresees something, it always foresees the worst. And its "worst" is as petty and mediocre as it is ugly—truly the most nauseating condition of human consciousness and matter. Well, I am right in the middle of this; I've been in it for months, and my way of being in it is to go through every possible illness.

> **65.247** — That material mind loves disasters and attracts them—it even creates them—because it needs the emotional stimulation to arouse its unconsciousness. Whatever is unconscious, whatever is inert needs violent emotions to shake itself and wake up. And that need creates a sort of attraction or morbid imagination for those things; it is forever contemplating every possible disaster or opening the door to negative suggestions. For the least little pain: "Oh, is it cancer?"

> **68.910** — There are whole worlds of suggestions. With one wave of suggestions, everything is terrifying; with another wave of suggestions, everything is delightful; with still another, everything is magnificent. . . .

63.38 — The physical substance, the very elementary consciousness that is in the physical substance, has been so mistreated that it finds it hard to believe that things could be other than what they are at present. I have this experience: when the concrete and utterly tangible intervention of the supreme Power, the supreme Light, is experienced by the physical substance, it's a new wonderment each time; and in that wonderment comes something like: "Is this really possible?" It is like a dog, you know, that has been beaten so often that it expects nothing but beatings. It's sad. And that physical substance has a kind of anxiety towards mental force; whenever a mental force manifests, it cries out, "Oh, no! Enough of that! Enough!" As if that were the cause of all its torment. It feels mental force as something hard, dry, rigid, and implacable—especially dry and empty, empty of the true vibration. It appears to be considered as the Enemy. This morning, there was a vision, a sense of the course that led from animal to man, then of the return state above the animal where life, action, and movement are not the product of the mind but of a force that is felt as a force of shadowless light, of pure light casting no shadow, and absolutely peaceful; and in that peace, so harmonious and so sweet . . . oh, there's supreme repose!

No longer the nostalgic return to the peace of the mineral, but cellular repose in the great expanse without walls.

"Liberation" is in the body.

64.710 — The great difficulty in matter is that material consciousness, meaning the mind in matter, was formed under the pressure of hardships—hardships, obstacles, suffering, struggle. It was "put together," so to speak, by those things, and they gave it almost an imprint of pessimism and defeatism, which is certainly the greatest obstacle. I am forever having to check, push aside, or convert some pessimism, doubt, or defeatist imagination. How many times has it happened that, in the middle of an acute pain, just as it feels increasingly unbearable, a tiny inner movement takes place in the cells: they send out their SOS . . . everything stops, the pain vanishes. The pain is replaced by a feeling of blissful well-being. But the first reaction of that stupid material consciousness is: "Ha, let's see how long it lasts!" Naturally, this movement ruins everything. Everything must be begun again.

58.105 — The moment the body becomes conscious, it is conscious of its own falsehood! It is conscious of this law, that law, a third law, a fourth law, a tenth law—everything is a "law." "We are subjected to physical laws: if you do this, such and such will result, such and such will happen, and so on and so forth. . . . Why, they exude from every pore! We must understand that IT IS NOT TRUE—it is not true; all that is an absolute falsehood. It's NOT TRUE! If only people had the experience I had a few days ago. . . .

Sometimes, the meshes of the web opened and another state filtered in, which seemed as miraculous as green pastures might to a man just escaped from a dungeon:

This experience is supreme knowledge in action, with total cancellation of all consequences past and future. . . .

And this is where we are left wide-eyed:

Each second has its eternity and its own law, which is a law of absolute truth.

And then the meshes close again.

65.107 and 48 — I can tell you that the mental distortions of doctors are frightful. They stick in your brain, remain there, and come back out ten years later. Doctors have a hypnotic power over the material consciousness that is rather frightening. The doctor crystallizes the illness, makes it concrete, hard; then he gets the credit for curing it—when he can.

60.2510 — I have observed and I have seen the power of thought over the body. It is tremendous! One cannot imagine how tremendous it is. Even a subconscious thought, and sometimes an unconscious one, has an effect and provokes incredible results. For the last two years I have been studying that in detail. It's incredible! Tiny mental or vital reactions that seem TOTALLY unimportant to our ordinary consciousness have an effect on the cells and can create physical disorders. Though I know for certain that if one can bring that whole mass of the physical mind under control, everything is POSSIBLE; one is in control. It isn't a question of Fate—something completely beyond our control—or a "law of Nature" we have no power over. For two years, I have accumulated experiences in the tiniest details, things that may seem utterly futile—you have to accept that, shun all notions of grandeur and appreciate that in the minute work of establishing the true attitude within a few cells lies the key to everything.

60.511 — I went down somewhere in the consciousness, in an area of the consciousness that lives in a state of apprehension, alarm, fear, anxiety—it's really and truly terrifying. And we carry that inside us! We don't notice it, but it's there. It's cowardly, and that's what can make you ill in a minute. It's in the cells' subconscious; that's where it's rooted. One has to go all the way down there to change it. And that does make for unpleasant moments, you know.

63.196 — It is as if the problem were becoming ever closer and pressing and crushing. It's that work in the physical mind, the material mind. I am seeking my way downward—a way out downward—and that's what I can't find. The way I am seeking is always down, down—never going up; always down, down. Oh! I have no idea when it will end!

60.1312 — It is a crass, seething mass. But how do you prevent that stupid, vulgar, and most of all defeatist automatic process from acting up constantly? It is really an automaton: no conscious will can move it, nothing. And it's closely connected with bodily illnesses. I am right in the middle of the problem.

Then the "problem" is unmasked, meaning the wall becomes clear, and the moment you know what the wall is, you begin to have the key. Oddly enough, Mother chanced upon the wall thanks to someone in her entourage afflicted with Parkinson's disease:

> **65.1812 and 63.1811** — When this material mind is seized with an idea, it is literally possessed by it and can hardly free itself from it. And that's what illnesses are. Take Parkinson's disease, for instance: that trembling is the result of being possessed by an idea, a state of hypnotism together with fear ingrained in matter. The two together: possession and fear. In the ancient scriptures, it was likened to a dog's crooked tail. It's really just that, a kind of WRINKLE that you try to flatten, and it comes right back, automatically and stupidly: you straighten it, and it wrinkles up again; you reject it, and it comes back again. It's extremely interesting, but quite sad. And ALL illnesses are like that, every single one, whatever their external appearance—the external appearance is a certain form of the SAME THING, because forms and appearances follow every possible pattern, and those that fall into similar patterns are classified by doctors as this or that disease. . . . AND THE CELLS OF THE BODY OBEY THAT MATERIAL MIND.

Mother had reached the bottom.

This discovery, which doesn't look like much, is quite formidable. We spend our time looking for answers right and left, in chromosomes and molecules and penicillin, in the bag of tricks of our science which codifies the walls of the prison—only to realize that they are nothing but the code of our own state of walled-in hypnotism. "Do you know that those walls consist of ten billion atoms per DNA molecule, and there are millions of billions of atoms per cubic inch of matter—as many as there are grains of sand in all the oceans combined—and twenty kinds of amino acids and five sorts of nucleotides?"[1] So

1. This science is borrowed from the remarkable book of Dr. Robert Jastrow, *Red Giants and White Dwarfs* (New York: Warner Books, 1980).

how do you propose to get out of this? And then—then it's nothing more than the paper-thin creation of our own conception of matter: that's *not* where the obstacle is; that's *not* the wall. The wall is in what we think it is. Illnesses are what we think they are. Death is what we think it is. And all the "laws" of our species are what our species thinks they are. A mind ingrained in matter.

Then one can begin to conceive of a way out after all.

6

THE BREAKTHROUGH

If we had only our strength, it would be practically impossible to get through the microscopic web of the physical mind. It is a rubber web: force it, and it shuts again; hit it, and it bounces back. We could do that for centuries—this is what insures our species' stability. But, occasionally, something very interesting happens: for a few seconds the mesh loosens. And that's when an overwhelming inrush occurs—something *really* overwhelming—and we see immediately why it lasts only a few seconds: some adaptation is necessary. A carp suddenly plunged to a depth of six thousand feet is simply crushed. So those seconds persistently recur through the years until the body becomes accustomed to them. But if a first opening occurs it will be automatically and irrepressibly repeated, because nothing is more stubborn than matter. In fact, this descent into the physical mind is so suffocating that it causes an irresistible cry for air that eventually brings about the first inrush of the other "environment." This same law seems indeed to apply throughout all the species: a considerable degree of suffocation or destruction of the existing environment is necessary to enable another environment to manifest. The obstacle is the lever. Our present epoch remarkably resembles the end of the dinosaur era on an earth they had ravaged; we must find another way to live or breathe, or to stop suffocating. And each species has its pioneer, a first fish that experiments with pulmonary respiration or something else; someone who takes the first step. Sri Aurobindo and Mother are neither philosophers nor sages nor saints; they are the pioneers or experimenters of the next species.

The Supramental Vibration

The first time a tear occurred in the web was in 1958, the year of *Explorer I,* the first American satellite. It would happen again and again, in greater and greater doses, until the great exit toward the other state of 1962. Here Mother describes the experience, which is very similar each time:

> **58.811**—I was descending into something like a crevice between two steep rocks, rocks made of something harder than basalt, but metallic at the same time. It was endless and bottomless, growing narrower and narrower like a funnel. And the bottom was invisible—pitch black. I kept going down and down like that, without air, without light . . . suffocating. Suddenly, as if I had hit a spring at the very bottom—a spring I hadn't seen—I was hurled out of the crevice with tremendous force and cast into a formless, limitless vastness. It was all-powerful and infinitely rich, as if this vastness were made of countless imperceptible dots—dots taking up no space—of a warm, dark gold. All that was absolutely alive, alive with a power that appeared infinite. And yet motionless. A perfect immobility, but containing an incredible intensity of movement and life! A life that was so . . . multitudinous that you can only call it infinite. And an intensity, a power, a force, and a peace—the peace of eternity. Silence. Calm. A POWER capable of everything. Everything. There was that whole impression of power, of warmth, of gold. It didn't feel fluid; it was like a powdering. And each of these "things" (I can't call them particles, or fragments, or even dots, unless we take dots in the mathematical sense of a point that takes up no space) was like living gold: a powdering of warm gold; I can't say they were bright or dark, nor was this a light; a multitude of tiny gold dots, and nothing but

that. With a fantastic self-contained power and warmth! And at the same time a feeling of plenitude, of the peace that stems from absolute power. It was movement at its utmost, infinitely faster than anything we can possibly imagine, and absolute peace, perfect tranquility at the same time.

Suddenly, it seemed that Mother had entered the atomic level, that her body was living quantum physics. Overwhelming motion in perfect stillness—such seems to be the constant characteristic of the experience. Then it recurred with greater precision and intensity.

58.169 — The other day it happened in my bathroom. It came, taking over the entire body. It went up like this; all the cells were trembling. With such power! So I let the experience develop, and the vibration kept amplifying, growing and growing, and all the cells of the body were seized with intense aspiration . . . as if the body itself were growing larger—it was becoming formidable. It felt as if everything were about to explode. But this has such transforming power! I felt that, if I continued, something would happen, in the sense that some equilibrium in the cells of the body would be altered. But this has a great action, a very great action: it can prevent an accident.

That is a mystery to which we will return once the experience reaches its full development.

58.115 — It has a peculiar cohesive effect: All the cellular life becomes one solid and compact mass of incredible concentration—with a SINGLE vibration. Instead of the body's many usual vibrations, there is only one single vibration. As if all the cells of the body were . . . a single mass.

61.241 — The entire body became a SINGLE, extremely rapid and intense vibration, but motionless. I don't know how to explain it because it wasn't moving space-wise, and yet it was a vibration (meaning, it wasn't immobile), but it was immobile in space. This was in the body, as if EACH cell had a vibration and there was but a single BLOCK of vibrations.

One cannot help thinking about the whirl of electrons around the nucleus, so fast it appears motionless, yet which gives matter its apparent solidity; also of the laser process in which the coherence of the light waves gives rise to an incredible amplification of power: "Instead of the body's many usual vibrations, there is only one single vibration."

63.185 — It was such a strong mass! It was much more solid than matter. It's something very particular and so solid! More solid, more material than matter. And it had such power, weight, density—incredible!

60.1110 — This extraordinary vibration . . . like a pulsation in the cells. During the first months, I was conscious, almost in every detail, of the myriads of cells that were opening up under this vibration.

That is the vibration Mother was to call the "supramental vibration"; physicists may have another name for it, but it is the same.

66.1511 — It's something that takes hold of the body: Such a warm, gentle vibration, and yet so terribly powerful at the same time!

64.253 — And that vibration feels like a fire. In fact, it is a vibration whose intensity is that of a higher fire. Several times the body has even felt it as the equivalent of a fever.

60.1211 — One must learn to widen, widen, not only the inner consciousness, but even this aggregate of cells—to widen this crystallization, as it were, to be capable of holding that force. I know. Two or three times, I felt the body was going to burst. I was about to say, "let's burst and be done with it." But then, weeks and sometimes months go by between one experience and another, so that some plasticity may come into these silly cells. A waste of time. Three times, however, I really thought I was on the verge of . . . coming apart. The first time, there was such a fever I was roasting from head to toe. Everything had turned golden red, and then . . . it was over.

72.151 — My body is living the process.

72.297 — As if to show you that, to conquer death, you must be ready to go through death. And it shows you there's only a difference, a tiny difference in attitude: The body can fall apart or it can become transformed, and it's . . . almost the same process.

And again the meshes close:

72.197 — The subconscious of the body is riddled with defeatism, and that must absolutely be changed. The subconscious has to be clarified to enable the new race to come into being—it's completely muddy. It is full of defeatism: the initial reaction is always defeatist. And that goes back to . . . A TREMENDOUS energy is obstructed by that ghastly thing.

Then the process of the crossing becomes clearer: one goes from the microscopic to the macroscopic, from the dust of atomic energy to the "undulation" of the other state:

63.35 — Now, the body feels this is not only a terrestrial movement but a universal movement, so incredibly rapid that it is imperceptible, beyond perception. It's like something that does not move space-wise but is beyond both immobility and movement, in the sense that its rapidity makes it absolutely imperceptible to all the senses. It is something new. While in that state, I have noticed that the rate of movement is greater than the force or power holding the cells together to make an individual form [hence Mother's fainting spells in the beginning]. And that state appears to be all-powerful. This must be the transition to the real thing. And that's constant; I constantly go from one state to the other: from here to there, here to there . . . so much so that sometimes—it is so strong—for a second or a minute, for some period of time, I don't know, you are neither this nor that; you feel there is nothing left anymore. It does not last. If it lasted, it would probably provoke a faint or something like that. But this is constant: from one to the other, one to the other, here to there. And between this and that, there is a transition . . . It's a strange kind of life, which is neither here nor there, which is not a mixture of both states, not their juxtaposition, but as if the two were operating through one another. It must be intracellular; that is, the mixture must be very microscopic, on the surface.

One crosses through the walls of the fishbowl, or through the wall of electrons, and it is exactly at that crossing, when the two states seem to operate simultaneously, or "through one another," as Mother says, that one grasps extraordinary secrets, which perhaps will be the fairy tale of the next species. Truly, we don't know if there has ever been a more fundamental event in the history of humanity than the experiment of Sri Aurobindo and Mother—our nuclear experiments seem like child's play in comparison, although scientific discoveries have certainly prepared us better to understand the present experiment.

Between Two States

This crossing through the wall or the web does not take place once and for all: That's it, we're out, in the other environment for good. If that were the case, the old body would probably die, having completed its evolutionary purpose, which was simply to take us over to the other state. The amphibian does not discard its old body; it acquires a new way of breathing with lungs that enables it to step into another state, in the open air, on the shores of this good earth, and gradually the very conditions of the new environment force it to develop new organs and a new way of being on earth. Mother's body stayed right here on this good earth, but the new shores looked a bit strange at first; they no longer had anything to do with the old retinal vision in the fishbowl; and the new conditions and the laws, if there were any, had to be explored—a sweeping change of "program." And as one does not land once and for all onto the new shore, since one occasionally falls back into the fishbowl (probably for the purpose of slow adaptation), what exactly triggers this falling back into the old state, and the return to the new one? What is the process of the transition? For years, Mother went back and forth between the two states, and it is precisely the moment of the transition, that hybrid state, one could say, which can enable us not only to explore the features and secrets of the new environment, but also to discover the reality of our *own environment,* which our physicists, biologists, and doctors think they have so thoroughly examined and codified. But their code is worthless! It is only suited to a certain thinking fishbowl. What there is, in fact, is a revolution whose scope we haven't begun to measure.

Here are the first stammerings of the new world:

> **61.66** — Take two sets of absolutely identical circumstances, not even a day apart, just a few hours apart: the same outer conditions and the same inner conditions. I mean, in both cases, the inner "mood" is the same, the circumstances of life are the same, the events are the same, and the people around me are pretty much the same. In one instance, the body (I mean the cellular consciousness) feels a sort of eurhythmy, an overall harmony in the atmosphere, as if everything dovetailed in a marvelous way, without conflict, without friction—everything flows and is organized in total harmony; everything is wonderful and the body feels well. And in the other instance . . . everything is the same,

the consciousness is the same, and this is where something escapes me, but that harmony is gone. For what reason? One simply does not know. Then the body starts to go wrong. Yet everything is identical—something eludes me; it's like trying to grasp something that keeps eluding you. What is it that eludes me? I just don't understand. What? More and more I feel—what? How can I explain that? A question of vibration in matter. It's incomprehensible. I mean, it is totally beyond all mental laws, all psychological laws: it's something self-existent. There certainly are a lot of question marks! The more you go into details, the more mysterious it becomes. It's almost like . . . standing on the edge between two worlds. It is the same world, and it's completely different. Is it two aspects of this world? I can't even say that. Yet it is the SAME world. . . .

The amphibian, too, landed in the same world; it wasn't a different earth.

And it is so subtle: if you do this *(Mother tilts her hand slightly to the right),* it's perfectly harmonious; if you do that *(she tilts it to the left),* everything is absurd, meaningless, laborious and painful. And it's the SAME thing! Everything is the same. If I step back and used big words, I could say that all this *(tilt to the right)* is truth, and all that *(to the left)* is falsehood—yet it's the SAME thing! In one case, you feel carried (not only the body, but the whole world and every circumstance), carried, floating in a beatific light; and in the other, it's oppressive, heavy, painful—EXACTLY the same thing, almost the same material vibrations! What is it? If we could put our finger on that, perhaps we would have everything—the whole secret. This must be the way truth became falsehood. But "the way"—what is this "way," precisely? What is the process? It is double. . . . It is double. And there is a kind of presentiment that the body alone can know the answer; that's what is so extraordinary!

And close to end of the experience, years later:

70.184 — Never, ever have I lived so totally and consciously in

> the other state, and it lasted two hours. Things were as real, as precise as they are here, which means that I don't know what the difference is. It's quite a . . . subtle difference; you don't feel there is a marked or heavy contrast; it's very slight. Quite remarkably, I couldn't have said: "This, here, is the subtle physical [the other state], and that is the material physical." They were . . . amazingly WITHIN ONE ANOTHER. They did not feel like TWO things, and yet they were very different—it might be a modality rather than a difference. I don't know how to say it.

Like the first bird realizing that it is not flying over another "subtle" world but over the same earth with another life mode. And Mother adds this, which conveys the full scope of the experience:

> I remember last night, I suddenly saw how a certain process works and I said to myself, "Ah, if only we knew that! How many things, how many fears, how many combinations would crumble, become obsolete." What appears to us as "the laws of Nature" or "inescapable principles" looked so absurd, so utterly ridiculous! With the true consciousness, it all crumbles. It's *we* who decide that it's "inescapable"! That's probably a . . . there's a POSITION, a position of consciousness that must be changed.

One side of the web versus the other.

There are thousands of fascinating experiences, and volumes would be needed to describe them (in fact, there are thirteen volumes, each four to six hundred pages long, making up *Mother's Agenda).* Here we can only give so many landmarks. But the essential fact is that from one side of the web of the physical mind to the other, the physical and physiological laws are not the same. And the other side is not far; it is just underneath that unrelenting whisper deep in the body.

> **73.173** — It is so different that I wonder . . . sometimes I wonder

how it is possible! At times, it is so new and unexpected, it's almost painful.

(Question:) You mean that you don't really leave matter?

No, not at all!

Is it a new state IN matter?

Yes. Yes, exactly. And it is ruled by something other than the sun; I don't know what. Probably the supramental consciousness.

70.129 — You see, I feel I am right in the middle of a world I know nothing about, struggling with laws I know nothing about, in order to make a change I know nothing about. What is the nature of that change?

Yes, but, Mother, I have a strong feeling that through that obscurity and that ignorance of the "laws," you are deliberately led to the point where the solution will emerge.

You are right. In a way, I could say that I think that too, although I don't "think," but . . . it's everything in between!

It cannot possibly fail!

Why?

Because you are the body of the world! Because it's really hope.

Isn't that just poetry?

Of course not! That's how it is. It's clear: the outer world is more and more hellish.

Yes, that's true.

Well, that's what's inside your body.

Someone has to take the first step.
Several persistent features stand out, however:

68.412 — All the time—it's really constant—the body has this same experience: in this position *(Mother turns her hand slightly* *to the right),* everything works out miraculously, miraculously, unbelievably; and it takes just a slight movement *(she turns her hand slightly to the left),* and everything is disgusting, goes wrong, and starts grating: just a slight movement. Then, again, things become miraculously wonderful. It becomes miraculous for microscopic, "unimportant" things, in other words, for EVERYTHING, without distinction of "important" and "unimportant"; and yet the thing itself is the SAME! But in one instance, you are in pain, you suffer, you feel miserable; and in the other . . . And it's the same thing. For example, the body feels this: it is really in a bad way, it has a cold, a pain here, a pain there; but when it takes a certain attitude, everything goes away! It doesn't exist; there is no trace of it left: no more cold, no more pain, nothing, all gone! Though it may well return [if one falls back into the other position]. Not only is it gone, but the surrounding CIRCUMSTANCES have changed too! In one case, everything is hostile, awry and in the other . . . And it takes no time at all; it isn't a "long process" of transformation; it's like something that suddenly flips over: flip-flop! *(Mother turns her hand to the left and to the right.).* This is like a concrete demonstration of the intervention of that wonderful consciousness in which all those things vanish as if they had no consistency, no reality, and they simply vanish. And it's a demonstration that this isn't just imagination but actual FACT—a demonstration of the power capable of changing all this present futile life into a marvel, just like that, with that simple reversal. Another way to put it would be to say this: the body has the feeling of being shut up in something—yes, shut up—shut up as if inside a box, but it sees through it; it sees and can also have some limited action through something that is still there but must disappear. So it keeps pushing and pushing to get to the secret; it's about to grasp it and . . .

69.315 — I have had the exact same experience Siddhartha Buddha had, but IN THE BODY. He had said Nirvana is the only way out; and at the same time, I had the state of true consciousness: his solution and the true solution. It was really interesting how the Buddhist solution is only a FIRST step, and how the true solution lies beyond it. Indeed, what is this creation? Separateness, meanness, and cruelty, and then suffering, decay, and disease, death and destruction; all that is part of the same thing. Well, the experience I had was the IRREALITY of those things, as if we had entered an irreal Falsehood, and everything disappears when we get out of it—it DOESN'T EXIST anymore, it no longer is. That's what is so frightening! All those things that are so real, so concrete, so terrifying for us do not exist! We've just entered a Falsehood. Why? How? What?

This "irreal Falsehood" is the very definition of the mental fishbowl. Of course, breathing with gills wasn't "false," but when you do find out about fresh air and pulmonary breathing, it's something else! And Mother added this:

And all the means—which we could call artificial, including Nirvana—all the means of getting out are worthless. I don't know, but salvation is PHYSICAL; not mental at all, but physical. I mean, it isn't found in flight; it's HERE. And it isn't veiled or hidden or anything; it's HERE. What is the part of the whole that prevents us from experiencing "that"? I don't know. It's here. It's HERE. And everything else, including death, becomes truly a falsehood, that is, something that doesn't exist.

But the old state does not dissolve all at once. It's as if one had to stay in it in order to dissolve it from within, or to infiltrate the new vibrational state into it.

67.197 — The millennial habit of being otherwise is so embedded that it feels like . . . pulling on a rubber band: The effect lasts as long as you pull on it, but as soon as you let go, even for a second, it snaps back to the way it was. Once the other movement is established, it will be natural; that tension won't be necessary anymore. But what an extraordinary feeling, this irreality of pain, this irreality of illness, this irreality . . . There are moments of indescribable glory, you know. But the other is there, too, all around, pressing.

> **68.49** — The entire material creation is under a fabric—a fabric we might call "catastrophic"—of ill-wills. It's like a web, yes, a defeatist, catastrophic web, in which anything you attempt will fail, and which is filled with every possible accident and every ill-will. It's like a web. And the body is being taught to get out of this. It's literally blended in with the force of materialization and expression; it's something that is blended with the material creation. It is the cause of illness, the cause of accidents—the cause of all destructive things.

Then the vibrational quality of the two states becomes more distinct:

> **62.412** — The quality of these two vibrations, which are still separate so one can be conscious of them both, is simply indescribable! One is fragmentation—infinite fragmentation—and utter instability, while the other is eternal stillness, infinite vastness of absolute light. The consciousness still goes from one to the other.

> **69.304** — I am being given a sort of demonstration. Man attaches a great importance to life and to death; for him they're quite different, and death is a rather major event(!) But I am being shown to what extent the disruption that results materially in what people call "death" is, in fact, ever present with the other: The all-encompassing harmony [the other state], which is the very essence of life, is together with the division, the fragmentation—which is superficial, irreal, whose existence is artificial—that causes death; how the two states are so closely fused with one another that one can go from one to the other at any moment and in any circumstance. It's not what people think, that something "serious" is required; it's just being here or being there *(slight movement to the right or the left)*, and that's all. Being here *(to the left)* and remaining here means death; and being there *(to the right)* is eternal life, absolute power; and you can't even call it peace, it's . . . something immutable. But both are there: that state and this state are there together.

65.2311 and 63.78 — You know what it's like to feel really uncomfortable, breathing with difficulty, feeling nauseous, helpless, incapable of moving or thinking or anything; then, suddenly, consciousness—the corporeal consciousness of the vibration of love, which is the very essence of creation. Only one second: everything brightens; gone, everything is gone! So you look at yourself in disbelief—everything is gone! It's just like turning a prism upside down: everything vanishes at once. The only thing left is that stupid habit the body has of remembering. And, of course, as it remembers . . . In one case, there is a sort of inner silence in the cells, a deep peace that doesn't prevent movement, even rapid movement—it simply takes place on a foundation of eternal peace; and in the other case, there's that inner restlessness, that fidgeting.

The very definition of the physical mind.

61.26 — I go off into the experience, and ten minutes later I realize I was in that state with my pen in midair! I have had instances such as these when you no longer understand anything, you no longer know anything, think anything, want anything, or can do anything; you are just . . . stopped. And then I see the people around me, looking at me and saying to themselves, "Oh, Mother is dotty."

69.1810 — The body has a feeling that the highest vibration, the vibration of the true consciousness, is so intense that it is the equivalent of inertia, of immobility; its intensity is imperceptible (for us). That intensity is so great that, for us, it amounts to inertia. And it is a state of immortality, immutably peaceful and still, amidst something resembling waves of overwhelming speed, so fast that they seem motionless. That is how it is: nothing moves (in appearance) in a fantastic movement. Yet it seems so natural, so simple! Then when you return to this side . . . Truly, the ordinary state, the old state, is conscious death and suffering; while in the other state, death and suffering seem absolutely irreal! That's it.

It would seem that at the threshold of the body, just where that primal mind connects with bodily matter and trembles and contracts like Parkinson's disease, where it blends with the very whirl of the electrons in their ceaseless movement and with them forms a solid wall, some kind of reversal takes place: from "infinite fragmentation" in constant trembling to those "vibrations of overwhelming speed" in perfect immobility, like going from Newtonian physics to intergalactic physics, or perhaps even to a new physics.

63.232 — At any time at all, if I stop talking or listening or working, it's like great blissful wings, as vast as the world, beating slowly. There's that feeling of immense wings—not two: they are all around and extend everywhere.

72.315 — There is no longer a sense of time, as if another time had entered this one.

Another physics has entered matter.

7

THE NEW PHYSICS

The Other Time

These experiences of "the other side of the web" may seem purely subjective, with no material consequence for the old environment we live in. "Well, it's all very nice, this 'fantastic energy,' this 'irreality' of illness and death, but here, in the old fishbowl, we're still really sick and we keep on really dying." That's a "real" fact. But the extraordinary thing in Mother's experience—truly a breakthrough in the evolution of our species—is that, in fact, we are shut in a bowl of *physical* irreality. *Physical* laws are not as we think they are, *physical* illness and death are not as we think and feel them. All our sensations and perceptions of the physical world are false. Hence, we can get out of it *physically.* Leaving the false perception doesn't lead to Nirvana, to heaven, or to death; it leads to the true physical reality, true matter—matter as it is. To another life in matter. For Mother's story is not the story of some freak of nature going off into another state, like the amphibian leaving the old ocean of falsehood and irreality and emerging into the open air; it is two states or two worlds *within one another,* and going into the new state alters the *physical* laws of the old state. It is going from false matter to true matter, from the false laws to the true law of the world.

This is Mother's very first cry in 1958, when a tear occurred in the web. She was eighty years old in 1958, and her experience was to continue for another fifteen years.

> **58.105** — The moment you are in the other consciousness, all these things that seem so real and concrete change INSTANTLY! There are a number of physical conditions—physical—in my body that have changed instantly. It didn't last long enough for everything to change, but some things changed and have never reverted. Which means that if that consciousness were constant, it would be a fantastic and perpetual miracle! But from the supramental point of view, it would not be a miracle at all, it would be perfectly normal.

Indeed, there is nothing "miraculous" to it—no more miraculous than Newton's apple falling to the ground at a certain speed. But this is just the point: *at a certain speed,* i.e. with respect to a certain frame of reference. And here is where Mother's corporeal experiences have more to do with Einstein's physics.

One of Mother's first comments in 1962, and her first cry after the "great exit" from the web was: "Death is an illusion, sickness is an illusion, ignorance is an illusion . . . only love, and love, and love—immense, formidable, stupendous, carrying everything"; and this short passage contains in seed the "miracle" of true matter:

> **62.66** — The sense of time disappears completely in . . . it's an inner immobility. But a moving immobility!

And with her usual humor Mother added: "If it continues, they'll want to put me in a madhouse!" But it is we who are shut behind walls, for without a doubt these "vibrations of overwhelming speed"—which may be electromagnetic waves or perhaps those of the "unified field"—move so fast that they appear motionless. And we know from Einstein that a change of speed results in a change of time.

But let us hear Mother develop her experience in all directions, beginning with the first steps:

> **62.315** — Suddenly, for no apparent reason—I still haven't been able to figure out how or why—you FALL, as it were, into the other room [as she sometimes called the old human state], as if you had tripped; and it starts to hurt here and to hurt there—you are uncomfortable. Then suddenly, it's as if you moved to another room, crossed a threshold or a wall, automatically. Almost without noticing it, I find myself in a position in which everything flows like a river of quiet peace—it's truly marvelous: All the creation, all of life, all movements and all things are one single mass, as it were, and this body feels like a very homogeneous part of this whole; and everything flows like an infinite river of smiling peace. Then, bang! You trip again, and again you are SITUATED, you are in a certain place, at a certain MOMENT in time; and a pain here, a pain there, a pain . . .

One reenters Time, which is the time of pain and death.

> **62.1210** — This is getting very concrete: You do this *(gesture to the left)*, and everything becomes artificial, hard, dry, false, deceitful—artificial. You do that *(gesture to the right)*, and everything becomes vast, calm, luminous, limitless, happy. That's all it takes *(Mother tilts her hand from side to side)*. How? Where? It cannot be described, but it's only a movement of consciousness. And the difference between true consciousness and false consciousness is becoming increasingly precise and at the same time THIN: no "great" things are required to get out of it. It's like a very hard, thin little rind—very hard, though malleable, but very, very dry and very thin.

It is the wall of the fishbowl. And Mother adds this very revealing comment:

> **64.118** — There is like a film of difficulties and complications added on by the human consciousness. It's much more prominent

in man than in animals. Animals don't have that; it's inherent in man and in mental development, and it's very thin, thin and dry like an onion skin, yet it spoils everything. It's the stupid onion skin of human mentality. You know how terribly thin an onion skin is, and yet nothing gets through it.

61.210 — It's the consciousness that is false! When you are open and in contact with "that," the vibration gives you strength, energy; and if you're quiet enough, it even fills you with a great Joy—all that in the cells of the body. You fall back into the ordinary consciousness, and immediately, the same thing, without anything having changed, THE VERY SAME VIBRATION COMING FROM THE SAME SOURCE turns into pain, discomfort, and a kind of sensation of instability and decrepitude. . . . I repeated the experiment three or four times to be sure, and it was absolutely automatic, like a chemical reaction: same conditions, same results. I found that quite interesting.

Of course the same vibration! There are not fifty kinds of vibrations in the universe, there's only one, the one that carries the worlds and us with it. And as it goes through the walls of the fishbowl, this same vibration is refracted, distorted, falsified—it's death. It may be only a toothache, but it's death just the same! It's part of the same family, because all ills lead *there,* to their culmination. They are of the same family of mortal and false vibrations.

Then the experience becomes increasingly clear:

63.35 — I am beginning to feel that Movement in the cells of my body. It is a movement like an eternal vibration, with no beginning and no end—something from all eternity, for all eternity [like a sine wave]. And it has no division of time; it's only when it is projected onto a screen that it takes the division of time.

That screen is exactly the formula of our human "onion skin": the fishbowl.

> And that Movement is so all-encompassing—all-encompassing and constant—that it gives a feeling of immobility to a normal perception.

"It takes the division of time," and in the same meshes it takes pain and death.

> **71.2512** — I am more and more convinced that we have a way of receiving things and reacting to them that CREATES the difficulties. If one can remain constantly in that consciousness [the other state], there are no difficulties, yet things are the SAME. The world is the same—it is seen and felt in a totally different way. Take death, for instance; it's a transitory phenomenon, which to us seems to have been there forever; it's forever because our consciousness cuts everything up. But in that divine consciousness, oh! THINGS BECOME ALMOST INSTANTANEOUS, you understand? I can't explain it. It's hard to explain. It's like an object and its projection. Things ARE, but we see them as if projected onto a screen; they come one after another. Something like that. I feel I am on the way to discovering . . . what the illusion is that must be destroyed so that physical life can be uninterrupted—to discovering that death comes from a distortion of consciousness. That's it.

This is the whole transition from the fragmented vibration of our human consciousness, invested with time and death, to a stupendous vibration, so rapid it appears motionless, and invested with another kind of time.

With his equations of relativity, Einstein has shown that such "immutable" quantities as the mass of an object, the frequency of a vibration, and the time separating two physical events are related to the speed of the frame of reference within which the measurement takes place—in our case, the earth reference or that of our human fishbowl. Thus, a clock aboard a satellite at constant speed around the earth will count 60 seconds between two electronic bleeps, while an identical clock on earth will register 61 seconds between the same two signals: time "slows down" with speed. The greater the speed, the greater the slowing down. There is also the well-known story of the space traveler who returns to earth having aged less than his fellow humans who remained on earth. If the "frame of reference" approaches the speed of light, time vanishes, and all the laws of Newtonian physics break down altogether. As Mother put it, "Things become almost instantaneous." This is another "frame of reference," like Mother's body in those waves of overwhelming speed.

> **66.3112** — It's something different. . . . It's very particular, it's an innumerable present.

> **69.234** — I don't know what is happening; something is happening in the cells and . . . It's a state, a state of intense vibration, which gives simultaneously a feeling of omnipotence—even in this *(Mother points to her own body)*, in this old thing—a luminous omnipotence, and static; that is to say, the cells have a sense of eternity. Something totally new in the body, and which seems totally immobile. I don't know what it is; it isn't immobility, it isn't eternity. I don't know; it's "something" like that, simultaneously power, light, and true love. To the point that when you leave that state, you wonder if you still have the same shape!

> **71.189** — It is a curious experience. The body feels it no longer belongs to the old way of being, but it knows that it isn't yet in the new one . . . it is no longer mortal but it isn't immortal. It's very strange. And sometimes I go from the most terrible discomfort to . . . a marvel. Sometimes there isn't a word in my head, nothing; other times, I see and know what's happening everywhere. It's really strange.

Then the experience becomes ever more explicit, conveying fabulous implications; for if time vanishes for the material, bodily consciousness, then decay vanishes, "consequences" vanish with their string of illnesses, accidents, and death; each "second" is new, so to speak; each moment of the universe is as new as if it had just been created; each instant of man is open and free of any past and future. The future is wholly present at each "second." So where is the consequence of yesterday and of these eighty-seven years, which will never be 87 years + one day? That day is no more; today is a brand-new day of the earth.

> **61.254** — In the usual state of consciousness, one does something *for* something. For instance, all the Vedic rishis had a goal: for them the goal was to find immortality. At any level there is always a goal. We ourselves speak of "supramental realization." But not too long ago, I don't know what happened, something took hold of me; not a thought, not a sensation, but rather like a condition—the irreality of the goal. Not its irreality; its uselessness. Not even uselessness; the nonexistence of the goal. Now, there is a sort of absoluteness in every second, in every movement, from the most subtle and spiritual to the most material—the sense of sequence has disappeared. The sequence has disappeared; this is not the "cause" of that, this is not done "for" that; one isn't aiming at "that." It all seems . . . It's very peculiar. An innumerable, perpetual, and simultaneous absolute. The sense of connection is gone, the sense of cause and effect is gone. All that belongs to the world of time and space. Each—each what? I can't say "movement" or "state of consciousness" or "vibration"—all these notions still belong to our mode of perception—so I say "thing"; "thing" doesn't mean anything. Each "thing" bears in itself its own absolute law. There is total lack of cause and effect, goal, intention. That type of connection *(Mother makes a horizontal gesture)* doesn't exist: only this *(vertical gesture)*. Something that has neither cause nor effect nor continuation nor intention. Intention of what! It's like this *(same vertical movement)*.

A vertical time, new at every "second."

> **62.206** — In the true position, there is neither friction nor wear

and tear.

58.105 — Each second has its own eternity and its own law.

As if Mother's body were *living* at the speed of light. Then one can begin to see the earth's miracle take shape.

The Substitution of Vibrations

No, the immortality of this old body is not the "goal." That would not be worth the trouble. "Who would care to wear one coat for a hundred years or be confined in one narrow and changeless lodging unto a long eternity?" said Sri Aurobindo.[1] Obviously, that new consciousness must gradually alter the features of its body, change the present corporeal rigidity into a new suppleness, free it from its dependence on gross matter for nourishment, allow it to discover other sources of energy, etc. It will take a few centuries. In the meantime, we have to last to allow the new state, which Mother called "the state without death" (note the nuance), the necessary time to evolve the desired transformations in this old, transitory body. That is not where our problem lies; it is an evolutionary process that follows its more or less accelerated course. What concerns us is the acceleration itself, the actual driving force of the change.

1930 — The true change of consciousness is one that will change the PHYSICAL conditions of the world and make it an entirely new creation.

Mother said that in 1930. That new physics is what interests us. We could perhaps call it supramental physics. How does it work?

To begin with, the new state is powerfully contagious. That

1. *Thoughts & Aphorisms,* XVII, 124.

is its foremost feature. The first mental vibrations in the anthropoid were most likely very contagious, too; and we all know today the power of a thought wave across the world. But here, strangely enough (or not so strangely), it is a power of material contagion, as if the fact of *living* in the true state, or in true matter, one could say, had the capacity to change the laws of the false, illusion-ridden matter we live in—its entire "logical" sequence of cause and effect, which are only the cause and effect of a certain illusion. The first "law" of the new physics is that each second is new and bears its own law, which does not depend on anything "before" and has no consequence "after." But how can a "state of consciousness" be contagious, we materialists of the old matter may wonder? "Consciousness" is a highly subjective thing. Actually, it might be the supreme objectivity of the world, but we have not known anything about it until now, because the only "consciousness" we know about is what revolves in our heads. Yet there is consciousness in matter, and it is a state of consciousness *in* matter, a cellular state of consciousness; and nothing is more contagious than matter, because it is one and the same continuous thing from one end of the universe to the other. The only separation is in our heads.

It is best to let Mother express her first groping discoveries in this new physics, as early as 1958, when the web was first rent:

> **58.66** — During the entire time it actively lasted [the new experience], it was absolutely impossible to have the least disorder in the body, and not only in the body but in ALL SURROUNDING MATTER. It's as if all the objects obeyed, and without any need to "decide" to obey; it was automatic. . . .

It is no longer willpower that communicates orders to matter; it is matter itself that communicates, automatically.

There was a divine harmony in EVERYTHING. It happened in my bathroom upstairs, obviously to show it's in the most trivial things, in everything, continuously. If this condition is established permanently, there can NO LONGER be any illnesses, it's impossible. There can no longer be accidents, there can no longer be disorders, and everything—probably in a progressive way—is to be brought in harmony as it was in harmony: all the objects in the bathroom were filled with joyful enthusiasm; everything obeyed, everything! I truly felt it was a first experience, meaning something new on the earth. It is really a state of absolute omniscience and omnipotence in the body; and it modifies all the surrounding vibrations. . . . It is likely that the greatest resistance will come from the most conscious beings, because of the mind itself wanting things to continue according to their mode of ignorance. So-called inert matter is much more readily responsive; it doesn't resist. And I am convinced that in plants, for instance, or in animals, the response will be much swifter than in humans. It will be more difficult to have to deal with a very organized mind; people who live in a well-organized and crystallized mental consciousness are as hard as a rock. They stand fast. In all likelihood, according to my experience, what is "unconscious" will follow more easily. It was delightful to watch the water running out of the tap, the mouthwash in the bottle, the glass, the washcloth; it all looked so happy and receptive!

61.113 — As I was walking yesterday, I was walking in a sort of universe that was exclusively the divine [the other state]: one could touch it, feel it; it was inside, outside, everywhere. For three-quarters of an hour, nothing but "that." Well, I can assure you, at that moment there were NO more problems! And what simplicity! Nothing to think about, nothing to want, nothing to "decide"—being, being, BEING! Being in an infinite diversity of infinite oneness. Everything was there, but nothing was separate; everything was in movement and nothing moved.

61.3010 — This new creation is something denser, more compact than the physical world ["mass" increases with speed, said Einstein]. We always tend to think that it's more ethereal, but it isn't! The impression that atmosphere gives me is of something more compact, yet at the same time not heavy or thick. But solid! Oh, so cohesive, so MASSIVE; yet, I don't know, it's entirely different from what one would expect. You cannot imagine what it's like. Something compact and WITHOUT DIVISION.

66.221 — And it is a marvelous way of being, infinitely superior to anything we have here! Here, there is always something wrong—a pain here, a pain there, or something or other, and then difficult circumstances; all that . . . takes on a different coloration. Everything becomes light, as it were, light, malleable. All the hardness and rigidity—gone. It changes everything! Everything is changed! I was brushing my teeth, you know, I was rinsing my eyes, doing the most ordinary things; their very nature was changed! There was a conscious vibration in the eye that was being washed, in the toothbrush, in . . . All was different! Clearly, if one masters that state, one can change all the surrounding circumstances.

Suddenly, the experience assumes a dimension that gives one pause:

67.127 — For two or three seconds, suddenly, it's as if one held the key. And what is usually called miracles appear as the simplest things in the world: "Why, it's quite simple; you just have to do this!" And then it goes away. When it's there, it is so simple, so NATURAL. And absolutely all-powerful. For example, something new seems to be coming to me: the power to heal. But not in the way it is usually described! I do not feel at all I am "healing" somebody; it's . . . putting things back in order. It's not even that. It's a LITTLE SOMETHING THAT DISAPPEARS, and this little something is essentially Falsehood.

In other words, the sealed bowl of physical irreality we live in.

It's very interesting. It is basically what gives the ordinary human consciousness the sense of reality—that's what has to disappear. What we call "concrete," a "concrete reality"—yes, that which gives you the sense of a "real" existence—this is the sensation that must disappear and be replaced by . . . It is inexpressible; it's like a universal pulsation. It is simultaneously all-light, all-power, all-intensity of love and a plenitude! It is so full that nothing else can exist beside that. And when "that" is here, in the body, in the cells, then it is enough to direct "that" toward someone or something, and everything is immediately put back in its place. So, in ordinary terms, it "heals"; the illness is cured. No, it doesn't cure it, it cancels it! That's it, the illness is canceled, made IRREAL.

And this is where we are left wide-eyed.

You see, this isn't the action of a "higher force" *through* matter, into others; it is a direct action, from matter to matter. What people usually call "healing power" is a great mental or vital power IMPOSING ITSELF against the resistance of matter; that's not at all the case here! It is the contagion of a vibration. So it's irrevocable.

61.271 — That state is a sort of absolute. An absolute that not only doesn't have to overcome obstacles or resistance, but that automatically CANCELS the resistance.

And here is the last clue to the mystery:

67.153 — When you stir water, it's no longer transparent. You create movements and those movements prevent the water from being transparent; you can no longer see through it. The same thing applies to the body; when you are quiet and vast, everything becomes limpid. And through that limpidity, you can see very clearly, you decide very clearly; everything falls into place and things organize themselves, by themselves; you don't even need to intervene. How can I say this? The whole universe moves at a fantastic speed, in perfect immobility. The words sound stupid, but that can be felt, seen, and lived. A luminous immobility moving at a fantastically rapid speed. And in that immobility, there is perfect transparency, and the problem does not exist. The solution precedes the problem.

Illness, death, accidents do not exist, cannot exist: The solution precedes the problem, or prevents the problem from arising; the problem is canceled as if it had never existed except in our own deceptive consciousness. Indeed, "ills" are made irreal or stripped of their illusory existence; and all our besieged existence becomes a perpetual miracle. A corporeal limpidity where all that no longer exists, no longer is. "A little something that disappears."

66.318 — This body lived the truth several times this morning for a few seconds, which could have been an eternity. One doesn't know if it lasted long or not; all that is over. And it doesn't abolish anything; that's what is so marvelous! Everything is there, and nothing is abolished; I mean, it doesn't abolish anything of the world. You don't even have the feeling that Falsehood is abolished; it simply doesn't exist. It's a little something . . . that changes everything. That's how a dead person can come to life again—like that, through that change.

And finally, the picture became clear, and not only clear but promising and accessible to the humans that we are. That day, Mother held the key to the "little something" separating the two states: the old human state, which she calls here the state of imperfection, and the new state, which she calls the state of perfection. And these two states are not at an astronomical or transcendental distance from one another; they are here, together, within one another, on this earth.

64.1211 and 253 — Perfection is here always, coexisting with imperfection; perfection and imperfection are coexisting always, and not only simultaneously but IN THE SAME PLACE. I don't know how to say it *(here, Mother puts the palms of her hands together)*. Which means that, at any time and in any circumstances, you can attain perfection. It isn't something that must be

> acquired gradually, through a gradual progress; perfection is an absolute state you can attain at any time. And so the conclusion is very interesting. When truth manifests [the other state], the vibration of falsehood disappears; it is canceled out as if it had never existed before the vibration of truth replacing it. You see, truth is here, falsehood is here *(Mother puts her palms together);* perfection is here, imperfection is here. They coexist in the same place. The minute you perceive perfection, imperfection disappears, illusion disappears. In other words, the capacity to live and to be that true vibration seems to have the power to SUBSTITUTE that vibration for the vibration of falsehood to the extent that . . . For instance, say the outcome of the false vibration should be an accident or some catastrophe; but if among these vibrations is a consciousness capable of becoming conscious of the vibration of truth, it can, it MUST, cancel the other and prevent the disaster. . . . There's this growing feeling that the True is the only means of changing the world, that all the other processes of gradual transformation are always tangential—they come closer and closer but never make it—and the ultimate step must be this: the substitution of the true vibration.

The substitution of the new, supramental physics for the old mental, scientific, and mortal physics.

Could it be that one day, suddenly, throughout the world the vibration of truth might break through the mesh of our web and cancel out, make irreal the horror, the pain, the death we live in; and that we might awaken in a new world in which the old laws of death will not make any sense and will vanish like a futile dream? Not a gradual process, but a sudden change that would catch us so unprepared that we would drop our entire old arsenal, and find ourselves bursting with an immense laughter.

And the earth looks at itself as if for the first time.

Above all, we should not make the mistake of thinking that that experience is reserved for some rare and privileged human beings living in exceptional circumstances. It is an experience we *all* can have in a material and corporeal way, and many have it *without being aware of it*. For it seems so simple, so natural that one does not notice it. The trouble with the secret is that it is right under our nose.

The Transparent Secret

What is very difficult for us to grasp is that we are immersed from head to toe (especially the head) in a world of *physical* irreality. Despite ourselves, even though we may begin to understand the truth a little, our first spontaneous, automatic, bodily reaction is: "Come on, I *see* this, I can touch it, it's 'concrete'; come on, gravity is a fact, things do fall to the ground; come on now, this is AN ILLNESS—the doctor said so and everyone says so. Try to jump off a cliff and see what happens!" Indeed, let us hasten to say, the idea is *not* to be unreasonable with our old temporary laws; it is far more serious than that. We must understand *how the irreality works.*

We have already mentioned the microscopic web that pervades every gesture we make, every step we take, every reflex and every nerve of our body: "You can't and you mustn't, be careful of this and be careful of that; this is dangerous, this is fatal—and this is not possible, not possible. . . ." Everything is "not possible" to that apprehensive, disaster-oriented, defeatist creature. Here we exaggerate the phenomenon with words, but it is actually a tiny trepidation in matter, something like a microscopic fear, or a tetanus-like movement embedded in the bodily substance. It is probably the frightened memory of a little cell forced to separate and protect itself from an enormous, devouring, seething magma. And that continuous, infinitesimal contraction causes a sort of ultra-rapid, imperceptible trepidation, which creates a veritable wall around our body and is strangely reminiscent of the electromagnetic barrier created by the incessant revolutions of atomic particles. This is the trepidation of the physical mind, referred to earlier, which is like the disaster-ridden memory of the earth; all living matter has evolved from disaster to disaster. The novelty with man is that he has "mentalized," i.e., crystallized and codified the "disaster." He has given it a frightening hypnotic power. Even if things were possible, they would be impossible.

And yet, the truth of the world is that everything is possible.

But we have delegated to the Machine the power of overcoming our "impossibilities," instead of looking into ourselves for the key to the great Possible.

I will give only two examples from Mother's experiences, then one from my own, illustrating the transparent secret, which is the very secret of the great Possible. The first experience took place during a local riot against the Pondicherry ashram.

> **65.192 and 242** — I saw that bombardment of rocks and the flames reaching high into the sky; the whole sky was red. I was simply sitting at my table, eating my dinner, when the attack began. And just before the attack, this experience came, this consciousness [of the other state]: I was no longer this body, I was the earth—the consciousness of the earth's physical truth, to be precise—with a peace! a stillness unknown to physical life. The whole attack appeared as an absolute falsehood, with no element of truth behind it [i.e., the great illusion of the fishbowl], but at the same time, simultaneously (it can't be put into words, but simultaneously), I had a microscopic perception all over the city, and in particular here over the ashram, of all the points of falsehood that MADE THE CONTACT POSSIBLE—the exact vibration of falsehood in each person and each thing that made the contact possible. Which means that, had the consciousness [of the other state] that was here been collective, had we been able to receive it collectively, nothing would have touched us; the rocks

would have been thrown, but we wouldn't have been touched. For example, a big rock was hurled and hit my window; and just at that moment I saw, I saw the exact vibration of falsehood in the consciousness of someone present that had enabled that stone to hit that particular spot. And I had the same perception simultaneously, everywhere, over the entire city. . . . So now I know—I know in an absolute and unforgettable way—what the vibration of truth in the physical is, what condition of the physical is required in order to *be* the truth. It's something immutable, PHYSICALLY immobile (mentally, there's nothing to it; it's easy). It's like a physical magnet for the true physical vibrations; it doesn't use the agency of the mind or even the vital; it's physical, a kind of magnet to attract the physical truth. . . .

The "physical truth" is precisely that of the other state, where that whole riot has no reality, no inherent truth, and hence *no power*. And Mother adds:

The vibrations of falsehood are a trepidation-like movement in matter. I could see, as clearly as you see material objects, the vibration that PRODUCED THE CONTACT with all that falsehood, and THE Vibration that made any contact impossible, that prevented any physical contact. Since then, several people have told me about their personal experience. For instance, X went out to call the police, and he had to cross the courtyard—rocks were literally raining on you—so everyone shouted to him, "Go back in! Go back in! You're crazy!" But he crossed the courtyard anyway, and not a single rock hit him. He had the feeling it was impossible for them to hit him. It was like a demonstration of the difference of vibration between the two states: the vibration that responds to Falsehood and the other that does not offer any response, preventing ANY possibility of CONTACT; they are different worlds. It's a world of truth and the other is a world of falsehood. And that world of truth is PHYSICAL, material; it isn't up in the heavens. This is what must come forward and take the place of the other.

(Question:) Is this the "true physical" Sri Aurobindo spoke of?

Yes, the true physical.

A material world where accidents, illnesses, death cannot happen. It is that world that must replace ours—not through a miracle, but simply through a change of vibration in matter. The vibration of the true state cancels out all the false, illusory vibrations of the fishbowl. A riot is not an "illusion." It is tangible, concrete, and even striking, yet it is an illusion. There exists a vibratory state of matter, a true state, in which there is no possibility of being hit; there is no contact between the two, like two worlds within one another. A world of physical truth and a world of physical falsehood. A world of physical freedom and a world of physical enslavement. A world of physical laws and a world outside illusory laws, which can be striking or not, deadly or not, gravitational or not . . . depending on whether you are here or there. A different position in matter. Precisely our species' new position that is grounded neither in our illusory spiritualisms nor in our illusory materialisms.

The truth of matter is something else.

And here is the other example, taken from Mother's childhood:

63.93 — I was nine or ten, and I was running with some girlfriends in the forest of Fontainebleau [near Paris]. The forest is dense enough that one doesn't see very far ahead. Because of my fast pace, I didn't realize I had almost reached the road, which was about ten feet below the path we were running on. It was a paved road, freshly paved. My momentum was so great that I couldn't stop. Zoom! I went sailing through the air. I was ten at the most, with absolutely no idea of miracles or marvels, nothing; I was simply hurled into the air. And I felt something supporting me, and I was literally carried down onto the ground, onto the stones. I got up (it had all seemed quite natural to me): not a scratch, not a speck of dust, nothing, absolutely intact. Then everyone rushed down to see, and I said, "It's nothing! I'm fine."

> But I remembered that impression: as if something were supporting me; that's how slowly I fell. And there was material evidence; since I was intact, it wasn't an illusion; and the road was freshly paved (you know French flintstone?). The soul was very much alive at that time; with all its might it resisted the intrusion of the world's material logic. These things seemed quite natural to me. I simply thought, "No accident can happen to me."

What is rather remarkable is that years later, as Mother was telling me this story, she saw a parallel between the slow-motion falling that set her down gently on the flintstone pavement and the great movement of wings mentioned earlier: "Like great blissful wings, as vast as the world, moving slowly; not two: they are all around and extend everywhere."

A different vibratory state of matter that even cancels gravity. There are no laws! There is only what we think, though it is not intellectual thinking; it is a microscopic thought in matter. We do not know the true physical reality, true matter, the world's true nature. We know only our trepidation in matter, which makes the contact possible with all disasters and creates disasters, as if a death-oriented, "scientific" cocoon were covering us from head to toe; and the more scientific it is, the more impenetrable it is. One gets scientifically hit by rocks and one breaks a leg most scientifically: "Come on, it's concrete, it's tangible, it's real."

And Mother exclaimed:

> **55.1412** — The sublime state *is* the natural state! It's you who are constantly in a state that is not natural, not normal, which is a falsehood, a distortion.

Lastly, I will give an example from my own experience. It happened in the deserted canyons near Pondicherry. I was sitting quietly, when out of a hollow came three men. Instantly I knew: "They're coming to kill me." I stayed where I was, not moving. Strangely, without any effort or concentration, I suddenly felt as if emptied of myself, without any reaction, without fear, without anything—like a stone, but a conscious stone looking unconcerned at some show, just as one can be both witness and actor in a dream. Except for its neutrality, the feeling was not really that of a stone, but rather that of a body, my body, as something utterly transparent and null, and a little indistinct. There was no movement, not the least quiver or throb; but I had nothing to do with it; there was no "self-control" involved, no effort. Something had taken hold of me in a transparent immobility. The three men were standing there: two in front, one behind. I didn't move. They were talking among themselves. Then a kind of voice in me said: "Get up." I rose, with my back to the canyon. One of them took off my watch, no doubt to simulate a robbery. The man behind came in front of me. And I saw the killer raise his arm to push me into the canyon. I followed the movement of that arm; my eyes met the gold-colored eyes of the killer. He lowered his arm, hesitating, as if unsure of what to do, or exactly why he was there. It seemed that he, too, now watched the scene as if it did not make any sense, or perhaps he had forgotten what he had come for. He turned around, the others turned around, and they left. They started to run as if panic-stricken. Then my heart suddenly remembered it should have been frightened, that they had wanted to kill me, and it started pounding like mad.

The only thing I know is that, had there been the slightest effort on my part, the slightest contraction or reaction to push these men back, even an inner refusal, a mere "no" inside, they would have killed me instantly. The opposition thus raised would have met and challenged their vibration, and the reaction would have touched off the whole process. But there was nothing, not even a breath of reaction. I was like thin air, as it were; the other man's vibration passed through me like a breeze, unobstructed. Can you kill a breeze? Some kind of contact is necessary in order to kill; you have to have a "handle"; here, there was no handle, because there was nothing. And since there was nothing, there *was* nothing!

In other words, for five or six minutes, by some grace, my physical mind was stopped. And that is how all "miracles" happen. But the true miracle is the natural state.

This is the earth of the next species.

A transparent secret.

> **60.1510**—It's funny how a thing in itself doesn't exist for people! What matters to them is their attitude toward the thing, what they think of it. It's very funny! Each thing bears in itself its own truth—its absolute truth, so luminous, so clear—and if you get in touch with THAT, everything falls marvelously into place. But people are NOT in touch with "that"; they are always in touch THROUGH their thought—the thought they have of it, the feeling they have of it, or sometimes worse.

What remains now is to learn to practice the secret: How to cancel the physical mind to arrive at the natural state, the pure cell, without mental and catastrophic and scientific coating. This is truly the extraordinary discovery of Mother and Sri Aurobindo: the "mind of the cells," the greatest biological revolution since a first living particle began to struggle at the frontier of inanimate matter and life.

This is the second evolutionary transition, no longer from matter to life, but from life to something else, which Mother called "overlife," and which could just as well be called "overdeath," since it is neither life as we know it, nor the death that goes with it.

It's what Sri Aurobindo called "the life divine."

8

THE MIND OF THE CELLS

Under its fourfold web, the body experiences nothing of the world as it is. This world "as it is" is, in fact, the great mystery of evolution. We know it, and so do other species, through a certain type of vision, binocular or other, within a certain light-spectrum, and through certain functional devices—claws, fins, vibrating cilia or electron microscopes—which do not so much describe the environment as they do our own behavior in the environment, or our successive behaviors and perceptions of a mysterious something that we express in a Bactrian, Greek, Latin, or electronic language. But the environment is always seen "through" something. The only difference between man and other species is that man adds a touch of Greco-Latin arrogance to the process, and his particular fins—cerebral "fins," that is—have taken over everything, and so effectively cut off any other means of communication that he does not even know what a fish or a bird knows, or any of the other little animals know, which, despite their ignorance of linear algebra, live in perfect harmony with their environment. This mysterious "something" in which we are immersed reveals itself gradually to the explorer of the descent into the body as a kind of astonishing marvel, in which all the laws and codes and "logical" connections turn out to be just the laws, the codes, etc. of our own instruments of measurement and perception. An incredibly free universe, an "instant marvel," as Mother put it. This is the "new evolution" Sri Aurobindo announced at the turn of the century, in which we will have to learn to live and to handle that rather dizzying freedom, unless our explosive arsenal overtakes us and brings us back, once again, on this earth or another, to the state of little flagellate organisms ever in search of the same freedom and marvel. Evolution is very stubborn, and whether it takes place on this or some other planet makes very little difference. But perhaps we could understand the marvel and make it happen sooner?

All the same, scientists will say, the little fish and ladybugs and all the other evolutionary creatures do not have our advanced mathematics and other impediments you mention, but they are nonetheless in a cellular sense prisoners of their species: They swim, crawl, die, and produce identical baby fish as their molecules of deoxyribonucleic acid (sorry) dictate. You may cancel death, accidents, and gravity, and see thousands of miles away as you see in your room, but there will always be a little Homo sapiens cell—obeying what? You say the cells do not obey the genetic code, that they are "hypnotized" and manipulated by the physical mind; well, first, prove it, and then explain how a "lawless" cell is going to cope, how it is going to stay bound to the other cells, through what "device" since there is no more device! What force is going to hold everything together and keep your body from disintegrating into the cosmos?

This calls to mind Sri Aurobindo conjuring up an imaginary logician at the beginning of the history of the earth:

> *When only matter was there and there was no life, if [that logician had been] told that there would soon be life on earth embodied in matter, he would have cried out, "It is impossible, it cannot be done. What! this mass of electrons, gases, chemical elements, this heap of*

mud and water and stones and inert metals, how are you going to get life into that? Will the metal walk?"[1]

Are the cells going to deviate from the genetic program? Come now!

We do not know whether all scientists are like that logician, but it does seem that they are very attached to their prison. Perhaps they are even the wardens of the material prison, just as others guard the spiritual prison.

The Training of the Cells

As a matter of fact, Mother did find it very hard not to disintegrate into the cosmos:

> **62.121** — It's very difficult with this body! Very difficult to keep its cohesion, to prevent it from dissolving into surrounding matter.

The record of Mother's fainting spells, whenever she got out of the web of the physical mind, is highly instructive, if only to prove negatively the controlling, coordinating, and imprisoning power of that physical mind. It took Mother five years, from 1962 to 1967, to understand the mechanism:

> **67.2211 and 65.217** — It all began when the doctors declared that I was very ill [in 1962]. Because the body was completely emptied of its habits and energies, I could not take a step without fainting. The moment I stood up to walk, wham! Down I went! I had to be held up constantly. But I never lost consciousness for a minute; I would faint, but I was conscious. I saw my body; I knew I had fainted. I didn't lose consciousness and my body didn't lose consciousness. Well, now I understand; at the beginning I didn't understand. I kept remembering what Sri Aurobindo had

1. Unpublished letter.

> said: that the physical mind is good for nothing, and the only way is to get rid of it. It's very difficult to get rid of it, because it is so closely connected to the amalgam of the physical body and its present form that whenever I tried and a deeper consciousness [that of the other state] sought to manifest, it caused me to faint. I mean that uniting or merging with the other state without that physical mind (it was canceled out) caused fainting. I didn't know what to do.

The story of those first five years outside the web actually sounds like a nonstop illness, with countless heart problems as well, to enable Mother to find the key to the cells' functioning. If the cells are to function "purely," that is, without any adjunct or intrusion of factors foreign to their substance, the body must be completely emptied of its old habits, its old coatings; this is the descent through all the "layers," the intellectual mind, then the emotional, then the sensory mind, and finally the physical mind. To show the scope of the undertaking, even the "survival instinct," that first protective wall of the species, has to go.

> **65.259** — One must accept infirmity, and even accept to look like an imbecile; one must accept everything, and there isn't one person in fifty million who has the courage to do it. Many people have also gone off elsewhere, into other, more or less subtle worlds. There are millions of ways to escape, you see, but only one way to stay, and that is to have the courage and endurance, to accept all the appearances of infirmity, powerlessness, ignorance—the appearances of the very negation of truth. But if one doesn't accept that, nothing will ever change. As for those who want to go on being great, luminous, strong, powerful, and so on, well, let them stay where they are; they can't do anything for the earth.

That year, Mother was eighty-seven years old.

We will give a few pointers along the course of what could

be called "the training of the cells." The first difficulty, of course, is that the cells panic, as they no longer know what to obey. Obviously, the old system has to go before the new one can manifest.

72.175 — All the body functions are undergoing a "change of command." The functions that were going on naturally, in compliance with the forces of Nature, are suddenly all disrupted. They're withdrawn. To be replaced by . . . something I call the divine—maybe Sri Aurobindo called it the supramental—which is the realization of tomorrow (I don't know what else to call it). When everything is thoroughly dislocated, when things are really quite bad, then "that" consents to intervene. The transition is not pleasant, with sharp pains and . . . impossibility to absorb food, etc. Obviously, someone had to do it.

69.811 — The moment of the "change of command" is always difficult, and if one doesn't know, it can be mistaken for signs of illness. Because the cells don't know what to obey anymore. But external signs are deceptive. Moreover, the physical consciousness—the one that operates on the cells—is used to exertion, struggle, misery, defeat; it is so used to it that it's quite universal: the end, you know, the undisputed, centuries-old end—it weighs heavily. It's very difficult. It takes a very slow and very steady work to replace this sort of habit of defeat by something else.

63.91 — It is very difficult for the body to change, because it lives by sheer force of habit. Whenever a little of the true way of living infiltrates at a certain point, without thought, without reasoning, without anything resembling an idea, almost without sensation, almost automatically, the cells are seized by the panic of the unknown. There's instant panic at that particular POINT: one faints or one is on the verge of fainting or one has a terrible pain, or in any case something APPARENTLY goes wrong. So what to do? Wait patiently until that small number or large number of cells—that little corner of consciousness—learns its lesson. It may take one day, two days, then the "great" chaotic and traumatic event quiets down, becomes clear, and those cells begin to say to themselves: "God, how stupid we are!" It takes a little while, but they understand. But there are thousands upon thousands upon thousands of them!

According to the scientists, there are one hundred trillion cells in an adult human body.

64.1410 — The body is learning the "lesson of illness," of the illusion of illness. It's very, very funny to perceive the difference between what is really there, the reality of the disorder, whatever it is, and the old habit of feeling it and receiving it, the ordinary habit—what we call an illness—"I'm sick." It's very funny. In every case, if one keeps truly quiet (it's very easy at the vital and mental levels but a little difficult in the cells of the body; it requires some practice), if one succeeds in remaining truly calm, there always comes a little light, a warm little light, very bright and extraordinarily quiet, in the background, which seems to say, "You only have to will it." Then the cells panic: "What do you mean, 'will it'? How can we will anything! An illness is upon us—we're overwhelmed, it's an ILLNESS!" The whole comedy. Then something says, "Calm down, calm down, don't cling to your illness!" And then they accept. On THAT POINT, they accept; the next minute, the illness is over. Not even one minute; in just a few seconds, it's over. But then the cells remember: "What happened? I had a pain there!" Wham! Everything is back again. And the whole comedy repeats itself, continuously, in the same way. So if only they could learn the lesson once and for all. . . . Life is on the verge of becoming a marvel, but we don't know how to live it. We still need to learn.

63.277 — The biggest problem is that the body's very texture is made of ignorance, and so each time the force, the light, the power [of the other state] seek to penetrate somewhere, that ignorance has to be removed first. And each time, the same thing happens in every detail: a sort of refusal caused by ignorant stupidity. At every step, in each detail, it's always the same thing that has to be dissolved. The initial reaction is invariably a negation. Then there always comes a smile in response, and the pain disappears almost instantly; "that" sets in, luminous and calm. Note that it isn't final; it's only a first contact. The experience recurs on another occasion and this time meets with a beginning of cooperation; the cells have KNOWN that with "that," the condition was changed—interestingly, they remember—so they start cooperating, and the action is even swifter. Then a third time, several hours later, the pain returns, and the CELLS THEMSELVES call out, because they remember. So now I know the trick! This is all for the purpose of educating the cells, you see. It is not just a person who is ill and who must be completely cured; it's for the education of the cells, to teach them how to live.

70.283 — There is a fully conscious work, and I could even say "methodical," which is imposed upon the body so that one part after another, all the parts and all the groups of cells, may learn true life.

What is very interesting—and one might say, a moment of capital importance in cellular history—is that eventually the cells *themselves call out.* They awaken from their hypnotized inertia. Freed from its habits and coverings, the substance of the cells begins to reveal its true nature. Here, Mother notices some very interesting and novel features:

57.1710 — There are all kinds of freedom—mental freedom, vital freedom, spiritual freedom—which result from successive degrees of mastery. But there is a totally new freedom—that of the body. During the flu epidemic, for instance, I was living every day among people carrying flu germs. And one day, I clearly felt that the body had decided it wouldn't catch the flu. And it wasn't a "higher willpower" making the decision, you know; it was the body itself deciding. When you are high up in the consciousness, you see and know things, but after you've come back down into matter, it's like water disappearing into sand. Well, now things are different: the body has DIRECT control, without any outside intervention. It isn't a higher consciousness imposing itself on the body; it is the body itself awakening to freedom in its cells. It is a cellular freedom.

61.311 — I had a sort of perception of the almost total lack of relevance of the material, external appearance that expresses the body's condition. Whether the external, physical signs are this way or that way was a matter of total INDIFFERENCE to the consciousness of THE BODY. Suppose, for instance, that the body has some disorder—swollen legs or an upset stomach—well, that was of no importance whatsoever; IT DOES NOT HAVE THE LEAST EFFECT ON THE TRUE CONSCIOUSNESS OF THE BODY, whereas we tend to think that the body is very upset when it is sick or when something is wrong.

(Question:) But then, what is upset, if it isn't the body?

Oh! it's the physical mind, that stupid physical mind! That's what always makes all the fuss.

But what is it that suffers, then?

It's also through that physical mind, because the minute you calm down that creature, the pain stops! That's exactly what happened to me. The physical mind uses the nervous substance, you see; if you remove it from the nervous substance, you no longer feel anything! That is what causes the perception of sensations.

61.112 — As a matter of fact, the moment one completely gets out of the ordinary mind, no external sign is proof of anything, none whatsoever. One can't rely on anything, neither on good health and balance, nor on overall disorganization; it doesn't prove anything.

Suddenly, we feel like a child discovering a body that is no longer like anything he's known before; and yet, it is the true body, the true consciousness of the body. A mysterious stranger beneath its fourfold web.

62.1610 — Whenever I ask my body what IT wants, all the cells reply: "No! We are immortal, we want to be immortal. We're not tired, we are prepared to go on fighting for centuries if necessary!" And that's what I've noticed: the closer you get to the cell itself, the more it says: "Why, I am immortal!"

Then one reaches the heart of the secret:

64.710 — Recently, I experienced this: a sort of completely decentralized consciousness (I'm still speaking of physical consciousness), a decentralized consciousness that could be here or there, in this body or that body (what people call "this person" or "that person," though that notion no longer makes much sense), upon which a sort of universal consciousness came among the cells, as if asking the cells why they wanted to keep this present combination or this aggregate [Mother's present body], pointing out to them or making them feel the handicap resulting from the great number of years, for example, or from the external difficulties, or from all the deterioration caused by wear and tear. But it didn't make any difference to them! That universal consciousness was saying, "Look, here are the obstacles. . . ." And those obstacles were clearly seen: the mind's ingrained sort of pessimism. But the cells themselves couldn't care less! To them, it looked like an "accident" or some "unavoidable disease," but in any event something that WAS NOT PART OF THEIR NORMAL DEVELOPMENT and had been forced upon them: "We couldn't care less about that! . . ." From that moment on came a sort of LOWER power to act upon that physical mind, and that resulted in a MATERIAL power of separating oneself from it and of rejecting it. It was really as if something decisive had happened. It was followed by a movement of confident joy: "At last, we're free of this nightmare!" And then physical relief, as if breathing had become easier. It was like being shut up inside a shell—a suffocating shell—and suddenly an opening is made in the shell. And you can breathe. It was a totally physical, cellular action.

Mother then added the following, which opens rather astounding perspectives:

Indeed, the moment one goes down to that level—the level of

the cells, or even of the constitution of the cells—how much lighter it all is! That kind of heaviness of matter disappears; it starts being fluid and vibrant again, which would tend to prove that heaviness, thickness, inertia, immobility is something ADDED ON. It isn't an inherent quality of matter; it's false matter, that which we think or feel, but not matter itself, as it really is.

But then, if death, illnesses, accidents, pessimism, "inevitable defeat," gravity, and the rest of our "laws" are not part of the cell's normal development, what *is* the true cellular substance? What is a pure cell, as it really is? What is matter?

If all our gravitational forces break down, then what holds the cellular aggregate together?

The New Principle of Centralization

About four billion years ago, when a first living particle just emerged from inanimate matter, it did not have a memory, except the one connecting its atoms to the first hydrogen cloud. It quivered, throbbed, stretched itself to grow and absorb more matter, as the nucleus seeks to absorb its electrons, as galaxies seek to draw in other galaxies and the sun seeks to attract its planets. Already, that cell was seeking its universal totality, as if nothing could be unless it is everything, as if in the heart of things was a great memory for totality—hunger or love. A constant whirling about oneself aimed at encompassing more and more being and space, and at recovering the original oneness that had disintegrated in a burst of joy or love, or whatever it is that people put into equations but never into their lives. An infinitesimal movement that has gradually evolved its own laws, from habits and from the conditions of the environment, a first memory for living and repeating a useful or fruitful habit—a first habitual churning that would soon give rise to a trepidating and mortal cocoon, from which one had to get out to die and grow further. That was the first web: a tissue of crystallized habits. It is the one Mother was to encounter, only infinitely more complicated and solidified by the human mental habit. In short, at the "end" of evolution, the question was to know whether one could leave the cocoon without dying and join again with the universal totality imprinted in our atoms without losing the individuality painstakingly formed over billions of years of labor: to be both the part and the whole. The trouble was, that human, habit-bound formation, here referred to as the physical mind, was "so intimately connected to the amalgam of the body and its present form," said Mother, "that whenever I tried to get rid of it, I would simply faint." One scatters into the cosmos. Since undoing the habit meant undoing the person, a new principle of coagulation or centralization had to be found, which would no longer be based on the mechanical repetition of human habit. That was the mortal cocoon of all species, the web. Mother had clearly seen the problem:

> **69.1712** — Death is the decentralization of the consciousness contained in the cells of the body. The cells constituting the body are held in a certain form by a centralization of the consciousness within them, and, as long as that power of concentration prevails, the body cannot die. It is only when that power of concentration ceases that the cells are dispersed, and then the body dies. The very first step toward immortality is therefore to replace that mechanical concentration by a willful concentration.

And since there is no longer any intellectual, emotional, or sensory will—all habits long gone in the process of crossing through the layers—this new will has to be cellular, but a cellular will no longer based on habit—that would be going back to our mortal cocoon. Based on what, then?

During the "cellular training," the cells had gradually and

painfully learned that a "drop of that" can cure anything; they had learned to call "that," as the nucleus, perhaps, "learns" to snap up its electron. Yet, even in its most basic nature, a cell is very mechanical: it needs to repeat over and over again; in fact, it repeats immemorially all the stupidities of the human species, after doing the same for other species. Therefore, a different non-imprisoning kind of mechanism had to be found, one that would not churn out a new mortal cocoon around the cell, while still providing the necessary cohesion or centralization.

Mother found a way, a way so simple that it is accessible to everyone. With Mother, everything is simple. This way is not new, it is even quite ancient, but its application is new. It is what is called a *mantra* in India. It is the only mechanical means that Mother ever used.

Each thing, whether animate or inanimate, has its own particular vibration—a rock, a virus, fire, water, radium, anything at all. They are the vibrations of the force constituting the "object," its particular frequencies or wavelength, much like the quasars at the confines of our universe. It is the vibratory network that encloses the object and gives it a specific form. But a sound, even an inaudible sound, is also a vibration. As it happens, there is a very ancient science of sounds in India, a knowledge of the whole gamut of vibrations, from the most material objects to the highest states of consciousness. (States of consciousness also have a vibration: anger, joy, the fragrance of a plant, etc.; each has its own particular vibration or sound.) This often misused science is therefore used to recreate an object by producing its corresponding "sound": fire has a sound, water has a sound, anger has a sound, even supreme bliss has a sound. The adepts of this science usually use their knowledge for crude, profit-making purposes—for magic—which we need not go into. But there exist other sounds that have the power to evoke higher states of consciousness (poets know this), and as it is possible to incite anger in someone, it is just as possible to incite something else. Love too has a sound—it may even be *the* sound of the universe. Whatever the sound may be, it is a *mantra,* a vibration capable of recreating a certain state of consciousness (or, at the other extreme, a certain material state, though this may be the same thing). A mantra generally consists of one or several Sanskrit syllables.

Mother thus found her mantra.

From the beginning of this yoga in the body, she had clearly seen the repetitive powers of the cellular substance, and she had concluded that if one could establish another type of vibration in matter—one of sunshine, say, of light, expansive like love—instead of the usual egocentric, sordid, pessimistic, and mortal vibration, then perhaps one could impart this substance with a new principle of cohesion, no longer based on a mortal habit, but on a divine one. Instead of spinning a cocoon of death, the cells could spin eternal life. So Mother began repeating the mantra, her mantra, which for her evoked supreme love, which is supreme life. At first, the mantra, or vibration, is repeated in one's head, using one's mental will, then gradually it descends through all the levels of the being: in the heart, in the sensations, in the movements and even the memory of the body. Once it settles in the body, it never leaves it; it repeats itself as steadfastly as our "Oh, it's cancer," "Oh, it's gravity," "Oh, I have a pain!"—all the little oh's that make up an habit-bound, mortal body.

60.46 — Sound has a power in itself, and by making the body repeat a sound, one also makes it absorb the corresponding vibration. But the words must carry a life in themselves, a certain vibration (I don't mean an intellectual significance, nothing of the kind, but a vibration). And the effect on the body is extraordinary: it starts to vibrate and vibrate and vibrate. . . .

60.209 — I have seen that the mantra has an organizing effect on the subconscious, on the unconscious, on matter, on the cells, on all of that. It takes time, but through repetition and perseverance, it eventually works. It is similar to daily piano practice, for instance: You repeat mechanically, and eventually that fills your hands with consciousness; this fills the body with consciousness.

Thus, one begins to see what could be a new principle of centralization of the cells.

63.107 — It's like standing on the threshold of an extraordinary realization that depends on a tiny little thing.

Mother's mantra had seven syllables:

OM NAMO BHAGAVATEH

It is for all the seekers who would like to discover matter as it really is, without its false materialisms and the false spiritualisms that go with them—perhaps the spirit in the heart of matter.

Free Matter

These discoveries, amazing though they may be, are only the threshold of a new earth, as new as may have been the emergence of a first green prairie upon the rocky cover of this good earth, or the first human gaze upon that springtime of the earth. But a new gaze is not enough; we must learn to live with and to handle that fantastic cellular freedom. How does one go about it?

The last (and rather vertiginous) stage of the transition from one state to the other gives us the clue. Mother was no coward; at ninety, she was more youthful and adventurous than any child. In a way, she is the adventurer of the new species. No one in her entourage could make any sense of it; they thought she was old, incapacitated, senile. One wonders how the first reptile that suddenly begins growing wings must have felt.

Yet this very vertiginous unknown contained the key to the new functioning, as if to prove that the obstacle is always the lever:

> **63.207** — All the usual rhythms of the material world are changed. The body can no longer know in the same way it knew before. So there is a period in between: things are no longer this way, but they aren't yet that way; they are right in the middle. What makes it hard is that from all sides, incessantly, come pouring in all the stupid suggestions of people around me: old age, decrepitude, the possibility of death, illness, numbness, decrepitude. That pours in constantly, constantly, and so this poor body is constantly plagued.

The old species is all around. Discovering the new species is not enough; one must also not be killed by the old. A first anthropoid must have been very disturbing to the apes.

> **69.192** — The work is to change the conscious foundation of all the cells, but not all at once; that would be impossible. Even little by little, it's difficult enough. The moment that conscious foundation is altered, there is almost a kind of panic in the cells: "Oh! What's happening?" So sometimes it isn't easy. The cells are dealt with by groups or by body function (or part of one); and for some, it is a bit difficult. For a minute, there is almost a kind of anguish; you're sort of in midair, you know. It may last just a few seconds, but these few seconds are terrible. Yet even that comes from this silly survival instinct in all cellular consciousness. The body knows this; it knows it. It's an old habit. All the groups of cells, all the cellular organs must surrender totally, in total confidence; that's indispensable. For some, it comes spontaneously and naturally; others need a little nudging to learn to do it. So the various body functions are taken one by one, in a wonderfully logical sequence, in relation to the body's functioning. It is truly a wonder, though

the body remains quite a poor thing, to be sure. But then, there are all the worries around me, from anguish at the thought that the end is possible, to impatience for the end to come! A whole range of feelings from fear to eager desire: "Free at last!" Free at last to do all the stupidities I want to do! And the body is very sensitive to people's thoughts.

66.285 — I practically can't eat any more; I force myself, otherwise I would only take liquids. I feel as if I were groping in the dark and the slightest misstep could send me tumbling into the abyss. It is like standing on a ridge between two precipices. And this is something happening in the body's cells; it has nothing to do with psychology or even sensations.

71.2212 — At every minute, it's: "Do you want death, do you want life? Do you want death, do you want life? . . ."

69.1810 — Truly, the ordinary state, the old state is conscious death and suffering; while in the other state, death and suffering appear absolutely . . . irreal. That's all.

70.205 — Suddenly, the body finds itself outside all habits, all actions and reactions, all consequences, etc., and there's like a wonderment . . . then it disappears. It is so new for the material consciousness that, each time, one feels on the verge of—there's a minute of panic in the consciousness.

> *(Question:) I have often thought that if a caterpillar were suddenly given human eyes through accelerated evolution. . .*

Yes, that's it! You see, the body KNOWS it isn't sick, that this is no sickness, but an attempt at transformation; it knows it full well, but . . . there are all those centuries of habits.

And then this cry:

66.93 — What a bizarre condition! The slightest thing would make you lose contact. The body no longer depends on physical laws.

That "slightest thing" is death—death as known by *the old species.* It was necessary to reach the point where nothing is left of the old functioning. Indeed, one cannot be simultaneously a reptile and a bird; there comes a moment when one has to give way to let the other take off. And it is at *that* moment one seizes the key—*with one's body* (not with one's head, of course).

> **60.2611** — For three or four minutes, sometimes ten, I am dreadfully sick, with every symptom that this is the end. And it's just to make me find the—to make me have the experience, TO FIND THE STRENGTH. It is only at such "moments," when logically, according to ordinary physical logic, it's all over, that one grasps the key.

The key is extraordinarily simple: What is the reaction of a suffocating pair of lungs? They open the mouth and gasp for air. And what do the cells do when they suffocate, when they have no more support, no more habits to churn, when they are thrown into . . . nothing? They repeat the mantra. Instead of churning death, they begin churning the new life, the new vibration, the new force.

From layer to layer—thick, sticky, quavering—from the intellectual mind to the emotional mind to the sensory mind, the mantra bores down like a power-drill. It bores imperturbably, with all the virtue of an old chatterbox repeating the same thing over and over again, until it reaches the microscopic level of the physical mind. At that point, it is automatic: under pressure from the mantra, one mesh of the web gives way, causing instant panic, then another mesh and another mesh—lots of little educational fits of panic. Each time, an air hole opens in the web and the cells fasten on to what they can: the mantra. And then the process becomes extraordinarily interesting; it is contagious. Matter is a place of immediate contagion: nothing can stay separate or walled-off; things spread immediately, for the simple reason that matter is perfectly continuous, from one tiny cell to the farthest reaches of the universe.

67.28 — The energy had totally gone [Mother again had been gravely "ill"] to leave the body absolutely on its own, for its conversion, one could say. And I realized that that body consciousness was filled with the SAME aspiration, the SAME eagerness as the other parts of the being, only with a far greater stability than in any other part. There are no fluctuations as in the vital or in the mind; it's very stable, and it comes in pulsations, not far apart, starting with one detail, then spreading and becoming generalized.

63.36 — It's the mind of the cells that takes over a mantra and eventually repeats it automatically, and with such persistence! I have heard the cells repeat my mantra! It was like a chorus, each cell repeating automatically. It was like countless tiny little voices repeating the same sound over and over. They sounded like a church choir with many, many choir children—tiny little voices. But the sound was very distinct; I was amazed: the sound of the mantra.

67.2012 — Nowhere can a more unshakable resolve and aspiration be found than here *(Mother taps her body).* That is the very characteristic of matter. And when it has surrendered and has faith, the aspiration is absolutely stable and constant; it is something ESTABLISHED, and established effortlessly, spontaneously, naturally, once and for all. Thus, one can foresee that when matter becomes truly divine, its manifestation will be infinitely more complete, more perfect in its details, and more stable than anywhere else, in any other world.

58.115 — It's peculiar how the mantra has a cohesive effect: the entire cellular life becomes one solid and compact mass of incredible concentration, with a single vibration. Instead of the body's many usual vibrations, there is only one single vibration. It becomes as solid as a rock, in one single concentration, as if all the cells of the body were a single mass.

68.225 — All the time, all the time, even amid the worst difficulties, the mantra surges up from the cells like a golden hymn: the incantation, you know, the call.

Sri Aurobindo had discovered the same thing some forty years earlier, only he never explained his discovery, probably because explanations are useless: one must *become* in one's body. This is what he said:

And there is too an obscure mind of the body, of the very cells, molecules, corpuscles. Haeckel, the German materialist, spoke somewhere of the will in the atom, and recent science [Heisenberg], dealing with the incalculable individual variation in the activity of the electrons, comes near to perceiving that this is not a figure but the shadow thrown by a secret reality. This body-mind is a very tangible truth; owing to its obscurity and mechanical clinging to past movements and facile oblivion and rejection of the new, we find in it one of the chief obstacles to permeation by the supermind Force and the transformation of the functioning of the body. On the other hand, once effectively converted, it will be one of the most precious instruments for the stabilisation of the supramental Light and Force in material Nature.[1]

But for that, the old matter (or false matter, we should say) had to reach the point of suffocation.

And now, we face the question, the real question: what, indeed, is matter? Matter as it really is? True matter? We are told that it's this law + that law + this other law, and this amino acid + that nucleotide +. . . . Some infernal addition.

Yes, it is the addition of all the habits we have acquired to splash around in a first earthly culture medium. But "laws," there are none! There are only fossilized habits. One day in

1. *Letters on Yoga* XXII, 340.

1965, on some banal occasion, the whole picture became crystal clear. It concerned a disciple who had an incipient tumor on her neck:

> **65.266** — It is a tumor. Probably a hair that has curled inward, and that the body has covered with a layer of skin; then out of habit, it has continued to produce skin over it, layer after layer. It's a stupid "goodwill." And that's what happens in the case of most illnesses. . . .

That's what happens for everything in life and for all of matter: a stupid goodwill twisting itself in one direction or another according to the need of the moment. But what if there were a true need, a need for a true life? Mother added this:

> Remarkably, this is the origin of all habits. The cells feel: "This is what must be done, this is what must be done, this is . . ." *(Mother draws circles with her finger).* All chronic illnesses stem from that. There may be an accident—suppose something has happened, an accident—but then a sort of submissive and conscious goodwill causes it to repeat: "It must be repeated, it must be repeated . . ." *(Mother draws circles with her finger).* And it stops only if a consciousness in contact with the cells is able to make them understand: "No, in this case, you mustn't repeat!"

That consciousness in contact with the cells is the mantra. It is the undoing of habit. And thus, we understand that matter can become *anything.* There is total freedom of choice, provided we find the way to be in contact with the cells. Then Mother concluded:

> In some cases, that repetitive ability can be extremely useful. I even think that it is what gives a form its stability, otherwise we might change forms or appearances! Or become liquid.

We then understand that we are on the threshold of an

incredibly new life. The mantra is only a first step to help bore through the layers, while preventing the body from being scattered into a "frighteningly" free cosmos. The second step is to find how, with what instruments, we will be able to remold that free matter.

But matter *is free*.

The difficulty is that it is *overwhelmingly* free.

Once the fishbowl is shattered, there is an overwhelming inrush of the very energies that constitute matter and the worlds, which Sri Aurobindo called the "supramental force."

> **62.126** — It is an overwhelming power, so FREE, so independent of all circumstances, all reactions, all events. It's something else—something else entirely!

> **64.73** — A power that can crush everything and rebuild everything.

> **71.19** — The ordinary body consciousness is too thin, too fragile to bear this formidable power. And so this body is getting used to it. It feels as if it had suddenly caught a glimpse of an extraordinarily wonderful horizon, but overwhelmingly wonderful!

The dawn of a new life.

A New Mind

For a long time, I did not clearly understand the importance of that mind of the cells, except that, in a certain body called Mother, the old laws seemed to be losing their grip. I saw her go through one heart attack after another, with a smile, and all sorts of illnesses that would have devastated any strong man. I understood that her body was like a testing ground, and that once the mind of the cells has internalized the right vibrations of the mantra, the body can last virtually indefinitely. There was also that mysterious "other time," where accidents and all of life's troubles seemed to dissolve. All this could amount to a rather enviable and enchanted life compared to the one we know, but I still regarded it as an individual phenomenon, an exception—nothing radical with the power to alter the structure of the species as a whole.

Gradually, Mother opened my eyes.

> **71.1812** — You can't imagine how radical this is, my child! I could really say I have become a different person. Only this—the external appearance of the body—remains what it was. To what extent will it be able to change? Sri Aurobindo said that once the physical mind is transformed, the transformation of the body will follow NATURALLY. It is the consciousness that must change, the consciousness of the cells, you understand? That is a radical change. And there are no words to describe it, because it doesn't exist on earth; it was latent but never manifested.

Indeed, it was latent, since the same mind of the cells exists in animals. (According to Sri Aurobindo, there is even a type of mind in the atom.) It is this mind of the cells that quietly and harmoniously spins out every specie's habits, without the complications and crystallizations added on by our human physical mind. Hence, in her cellular substance, Mother had not only reached a pre-human stage, but, more radically still, the original condition of a first cell on earth, before it began spinning any habit. She was at the beginning of the world! Indeed, she had great difficulty not to disperse in the great culture medium. The initial reaction of any living substance is to protect itself, to build walls. The vibration of the mantra in each cell provided this protective wall, a network of vibrations dense enough to resist both the contagion of the environment and dispersion. And after that?

After comes the birth of a new species, as simply as that, *automatically.* But instead of an obscure, unconscious automatism that spins out habits as the result of bumping here or there, or because of the lack of food in a certain climatic zone—any of the "conditions" of the environment—it is a conscious automatism, that will gradually remold or refashion the nature of its body without falling prey to any habit, for it has none, or, shall we say, according to a new kind of habit, a new way of being in the world. In other words, a new species being slowly developed from within on the basis of the only mind it has left: the mind of the cells.

> **71.1812 and 65.218** — This body-mind, the only one I have left now, is undergoing a very rapid and interesting conversion. What could I call it? A transfer of power. The cells, all the material consciousness, used to obey the individual inner consciousness (in most cases the soul, or else the mind). But now, this material mind is in the process of setting up its own organization, as the other mind did, or, rather, as all the other forms of mind did, the mind of each level of being. It is getting an education, if you can imagine. It is learning things and organizing the ordinary knowledge of the material world. It's very interesting. You see, the memory that came from mental knowledge had disappeared long, long ago; I was receiving all necessary indications exclusively from above [the higher levels of consciousness]. But now, A SORT OF MEMORY IS BEING BUILT FROM BELOW. There has been a shift in the controlling will, as it were. It's no longer the same thing that makes you act; by "act," I mean everything: moving, walking, and so on. The greatest difficulty is with the nerves, because they are so used to the ordinary conscious will that when it stops and you want direct action instead, they go almost crazy. I experienced that yesterday morning, for over an hour; it was hard, but it taught me a lot. So all this is part of what can be called the "transfer of power." The old power withdraws. But before the body adapts to the new power, there is a critical interim period

> . . . and the minutes seem endless. But this mind of the cells, I assure you, is something totally new, totally new.

A new body being developed from below. But the process is so quiet, so invisible, it takes place through the development of so many imperceptible and microscopic new ways of being in the tiniest everyday movements and in the slightest nervous reactions that it is difficult to understand. And I didn't understand it very well. Mother tried to explain to me:

> **67.3012** — What the body is learning is this: to replace the mental rule of intelligence by the spiritual rule of consciousness [the other state]. And that makes a tremendous difference (though it doesn't look like much; you can't notice anything), to the point that it increases the body's capabilities a hundredfold. When the body is subjected to certain rules, however broad they may be, it is a slave to those rules, and its possibilities are limited accordingly. But when it is governed by the Spirit and Consciousness [of the other state], the possibilities and flexibility become exceptional! And this is how it will acquire the capacity to extend its life at will. The "necessities" have lost some of their authority; it becomes possible to go this way or that way. All the laws—the laws that were the laws of Nature—have lost some of their tyrannical power, you could say. It is like a progressive victory over all the "musts." Then all the laws of nature, all human laws, all habits, all the rules are gradually relaxed and eventually become nonexistent. Above all, that whole sense of rigidity, absoluteness and near-invincibility brought about by the mind will disappear.

But I still didn't understand the consequences of Mother's experience for the human species as a whole.

> **67.2211** — *(Question:) I understand what is happening in you, but—*
> But if it's happening in one body, it can happen in all bodies! I am not made of anything different from other human beings. My body is made in the same way, I eat the same things, and was

conceived in exactly the same way. And my body was as stupid, ignorant, unconscious and stubborn as every other body in the world. It all began when the doctors declared that I was very ill; that was the beginning of everything. Because the body was completely emptied of its habits and energies. Then very slowly, the cells awakened to a new receptivity. Otherwise it would be hopeless! Matter, you see, was originally less conscious than a rock—even a rock has already some form of organization—it was certainly worse than a rock: an inert, absolute unconscious. Then it gradually awakened. Well, it is the same here: for the animal to become a human being, all it took was the infusion of mental consciousness; and now, the consciousness that was buried deep, deep down is awakening. The mind has withdrawn, the vital has withdrawn (that's what gave the appearances of a very serious illness at the time), and in the body left to itself, the cells gradually began to awaken to consciousness. And from that, after further working and prodding (I don't know how long it will take), will emerge a new form, which will be what Sri Aurobindo called the supramental form. And it will be—whatever, I don't know what these new beings will be called. What will be their mode of expression? How will they communicate? In man, all that evolved very slowly. But when man emerged from the animal, there was no way to record the process. Now it's quite different, so it will be more interesting. . . .

Mother's Agenda is that process.

But even today, the overwhelming majority of humans and the human intellect are perfectly content with taking care of themselves and their little rounds of progress. They don't even want that there *be* anything else! Which means that the advent of the superhuman being may well go unnoticed, or misunderstood. It's hard to tell, since there is no precedent to compare it with, but if one of the great apes ever ran into the first man, it would more than likely have felt that he was a rather . . . strange being, and that's all. Man is used to thinking that anything higher than he has to be divine beings—that is, without a body—appearing in a burst of light. In other words, all the gods as they are conceived of. But it isn't like that at all!

Such is the present situation.

Are we going to continue searching for an answer in the "genetic program," which is only the program of the human habit, or will we go to the root of habits and unearth the cellular freedom and the power to remold our species?

Are we going to be blind to the process altogether, or simply let it unfold as usual, in spite of ourselves, with the usual shattering events of History, as it has always been up until now for every evolutionary transition from one species to another?

Because a little cell is highly contagious. The great whirlwind that has overtaken nations, continents, the human races with all their beliefs and unbeliefs, and every family and every little consciousness, may well signal the approach of the great evolutionary whirlwind that overtook the reptiles at the dawn of the mammals. We may not be so much in the twentieth century of a so-called Christian Era as in the thirty-five-millionth century since the appearance of a tiny unicellular organism.

Matter is highly contagious. We only know of contagion through reproduction or virus, but what do we know of the contagion or propagation of a vibration in Matter? It took nothing more than a mental vibration to make Einstein. And now, it's something else. Who wants something else? But whether we want it or not, IT WILL BE.

71.112 — It is almost like a new mind being formed.

70.143 — And the body is learning its lesson—all bodies, all bodies!

Not only are some twentieth-century or thirty-five-millionth-century cells surreptitiously dismantling the old habits of the world and infiltrating something so new that we do not even see or understand it, except through the disruption it creates, but there is also a new perception of the earth that is toppling our materialism and spiritualism and revealing something very strange—perhaps the true perception of the earth, without this side and that side, this life and that death—something Mother called "overlife," which we will try to describe:

> **61.273** — Yesterday I felt so strongly that all our constructions, all our habits, all our ways of seeing and reacting to things, all that was completely collapsing. I felt as if I was suspended in something totally different, something . . . I don't know. And really, the feeling that EVERYTHING we have lived, everything we have known, everything we have done is an utter illusion. When one has the spiritual realization that material life is an illusion (some people find it painful, but I found it so marvelously beautiful and joyous that it was one of the most beautiful experiences of my life); but here, it is the whole spiritual construction as one has lived it that is becoming a complete illusion, and not the same kind of illusion, but a far more serious illusion! And I am not exactly a baby; I have been doing yoga consciously for some sixty years.

> **71.0112** — It is a new mind. The way of perceiving time and space is becoming very different; it is changing completely. Where sight is concerned, for instance, I see more clearly with my eyes closed than with my eyes open, and yet it is the SAME vision! It's PHYSICAL VISION, purely physical, but a kind of physical that seems . . . fuller.

A new perception of the earth.

9

THE EYES OF THE BODY

We are before a great mystery.

I have been before that mystery for seven years, and sometimes I feel I understand, at other times everything vanishes. Yet, all the facts are there; I have thousands of experiences before my eyes. But then again, how can the caterpillar understand the dance of the butterfly over the pond? It is a very mysterious country—perhaps earthly, though who knows? And the thousands of experiences I have recorded are themselves very entangled and confusing (for me), because Mother did not land in that "country" all at once. Sometimes she saw it from afar, as with a bird's-eye view, across inner distances, and she would describe it and give it a name; at other times, the description would be different and so would the name, yet it was still the same country. But how does one know that? Ultimately, it was not "another" country at all but our very own; we were in it already. It does not seem to make sense. It is very hard for the caterpillar to make sense of the butterfly world—the butterfly looks quite mystical, and its pond, "supernatural." So the earth and its species go from one supernatural world to another, until they land in the great, ever-present natural world. Then "everything becomes self-evident," as Mother would say, though perhaps there will always remain a bit of "supernatural" ahead of us, and we will always be the past of a butterfly yet to be born: evolution is movement. It is very threatening to orthodoxies. Darwin did indeed commit "murder."

So let us go on with the murder.

The Net

57.107 — This is a perception or a sensation or an impression that is totally peculiar and new.

That was in 1957. Then four years later:

61.276 — I am just at the border, at the threshold. It's as if there were a semitransparent curtain and you see things on the other side; you try to grasp them, but you can't. Yet it feels so, so close! Sometimes, I suddenly see myself as a huge concentrated power, pushing and pushing with an inner concentration, to break through.

Then in 1964:

64.189 — I am on the threshold of a new perception of life. It's as if certain parts of the consciousness were changing from the caterpillar state to the butterfly state, or something like that.

And six years later, in 1970 (Mother was then ninety-two years old):

70.224 — There is a region with many scenes of nature, such as fields, gardens, etc., but they're all behind nets! There's a net of one color, a net of another color. . . . Absolutely everything is behind nets, as if one moved about with nets. But the net isn't fixed; the color and form of the net depends on what is behind it. And this is the means of communication. It's fortunate I don't speak to anyone, you know, because they would say I am losing my mind! And I see that WITH MY EYES OPEN, in broad daylight, if you can imagine! For instance, I am in my room—I am here, seeing people—and at the same time, I see one scene or another,

> shifting and moving, with that net between me and the scenes. The net seems to be—how can I put it?—what separates the true physical from the ordinary physical.

Since I was skeptical by nature, I often asked Mother, over the years, if this was not "a psychic's vision." But it wasn't. It is "the *same* vision, physical vision, purely physical, but a 'physical' that feels fuller." Furthermore, Mother was supposedly blind, so with what *physical* eyes was she seeing if it were not those of the ophthalmologist? It was obviously the "eyes" of the body, of the cells (I am reminded of some Russian experiments showing that a person is capable of distinguishing colors through the skin of the hands or even the skin of the stomach), but it is not as if little cellular eyes were watching a show. It is not a "vision"; it is *better* than a vision:

> **70.257** — Now, the body itself has the experience and it is MUCH TRUER. There is an intellectual attitude that places a kind of veil or . . . I don't know, something . . . something unreal over our perception of things. It's as if we were seeing THROUGH a certain veil or a certain atmosphere, whereas the body feels directly; it BECOMES. It feels things in itself. Instead of the experience being scaled down to the size of the individual, the individual widens to the size of the experience.

One cannot help thinking that those scenes behind nets are what the body sees through the web of the physical mind, until there is no more "through." Is it a coincidence that a researcher at the University of California,[1] using photographs taken with an electron microscope capable of distinguishing two dots one thousandth of a micron apart (.000 000 039 inches), recently observed: "One of the intracellular features of greatest interest to scientists is the network that looks like a fishnet draped throughout the main bodies—the cytoplasms—of the cells.

1 Dr. John E. Heuser, 1979.

Before discovery of the nets, the cell cytoplasm was thought to be more or less like jelly with no internal structure. Now scientists believe the networks help maintain the cell's shape"?

But the question is even more radical than a mere change of vision. One of the very first times Mother caught a glimpse of the other side of the web, of the other state, which she also called with Sri Aurobindo "the truth-consciousness" (meaning the consciousness of the truth of the world, as it really is), she made the following remark, which shows the full scope of the problem:

> **61.187** — There's like a veil of falsehood over truth; that's what is responsible for everything we see here. If that is removed, things will be totally different, totally. When you leave this ordinary consciousness and enter the truth-consciousness, you actually wonder how there can be such things as pain, misery, death and all that; there's a kind of amazement. You don't understand how that can happen, once you are on the other side. Though this experience is usually associated with the experience of irreality of the world as we know it, it's actually the irreality of the falsehood, not the irreality of the world!

Truly—and it is hardly a metaphor—we can say that we are shut in a physical glass bowl with a refractive index that causes all the misery and death and irreal falsehood of this world. If that refraction ceases, everything changes *physically*. But Mother adds the following, which was to become the main question over the years:

> That new consciousness most likely will become a permanent condition, but then a problem arises: How does one keep in contact with the world as it is in its distortion? Because I have noticed one thing: When that state is very strong in me, strong enough to resist everything that harasses it from the outside, if I say something, people understand nothing, nothing. Therefore, it must cancel some useful channel of contact. What would a

small supramental creation on earth be like? Is it conceivable? How would the connection between those beings and the ordinary world work?

Then, in 1968, came the second radical exit from the web. Once again, Mother almost died from it. A few days later, she tried to tell me what was happening, or what had happened (or what is going to happen, since tenses also seem to jump around when crossing the web's mesh):

68.288 — I am convinced the movement has begun. How long will it take to arrive at a concrete, visible, and organized realization? I don't know. Something has begun. It seems there will be an onrush of the new species, the new creation, or at any rate a new creation. A reorganization of the earth and a new creation. There was a moment when things were so acute . . . I usually don't lose my patience, but it had come to a point where absolutely everything was as if abolished in my being. Not only was I unable to speak, but my head felt as never before in my whole life—in pain, you know. I couldn't see anything, I couldn't hear anything. . . . But two or three times, I had moments, absolutely marvelous and unique moments—indescribable. It's indescribable. And scenes! Scenes of construction: huge cities under construction. Yes, the future world being built. I no longer heard, no longer saw, no longer spoke; I was in that all the time, night and day. A body without a mind or a vital—just those perceptions. The mind and the vital have been instruments to knead matter, to knead it over and over in every possible way: the vital through sensations and the mind through thoughts. But they seem to me to be temporary instruments that will be replaced by other states of consciousness. They represent a phase in the universal evolution, you see, and they will fall away when they become obsolete as instruments. And so I had the concrete experience of what matter is, kneaded as it is by the vital and the mind, but WITHOUT any vital or mind; it's something else altogether! I've had moments . . . Anything we can humanly feel or see is nothing compared to that. There were moments . . . absolutely marvelous moments.

But without a thought. And it wasn't "seen" as one sees a picture; it's BEING IN IT, being in a certain place. I have never seen or felt anything as beautiful as that, and it wasn't "felt." I don't know how to explain it. The body felt almost porous amid this—porous, without any resistance, as if it all went through it. I have spent hours . . . the most wonderful hours ever possible on earth. One night (just to show you how upside-down everything was), I had a rather sharp pain; I remained concentrated, and it seemed the night passed in a few minutes. And other days, I would be concentrated and now and then would ask the time, thinking I had been like that for hours and hours, and it had only been five minutes. You see, everything was—I can't say upside-down, but in a totally different order.

This is when clock time would disappear into the "immobility in movement" mentioned earlier.

72.2312 — It is the sense of time I don't understand. . . . I feel, I know that my body is being accustomed to something else.

66.3112 — Time no longer has the same reality. It's something else. It's very particular, an innumerable present.

69.127 — And then I go to America, I go to Europe, I go . . . all the time. I go to places in India. And all that is work, work, work, but so vivid! And with such humor! Here, things are always covered in many clothes; it's never the exact thing, but *there* it is the exact thing. It's very interesting, you know—life stripped of its deceptive appearances. People are used to disguising everything; there, that's all gone!

72.76 — It's something that the cells still don't understand, but they know, they feel. They feel as if they were thrust by force into a new world.

73.82 — It isn't going off into some inaccessible planes; it's RIGHT HERE. Only, for the moment, all the old habits and the general unconsciousness put a sort of cover over it, which is preventing us from seeing and feeling. That must be removed. And it's everywhere, you see, everywhere, all the time. It doesn't come and go; it's there all the time, everywhere. It's us; it's our stupidity that prevents us from feeling it. There's no need to go off at all, no need whatsoever.

72.275 — *(Question:) Where do you go when you suddenly go off like that?*

But I don't "go off"! I don't leave material life, only it appears different, as if it were made of something different.

This is a brief review of some of the main features, except for one we will soon discuss and which opens up rather peculiar perspectives. The essential fact, though, is that behind our "cover," or our "veil of irreality," there is a physical world invested with another innumerable and instantaneous vision, and another "vertical" time in which illnesses, accidents, and death *cannot* be. "The solution precedes the problem," as Mother would say. And yet, that other time is *physical;* when you are in it, you cannot be murdered in the canyons of Pondicherry (among other things).

Therefore, truly, "salvation is physical"; there is no need to run off to other "spiritual" worlds. Redemption is on this earth and in this body. We only need to get out of the net.

But can one do it alone?

Evolution means the whole earth.

The Living and the Dead

I must admit, I do not clearly understand this last, peculiar feature, but such is the fact. It began in 1959, nine years after the passing of Sri Aurobindo, as Mother was already struggling within the last layer of the physical mind, with occasionally odd little tears. One fine day in July, while boring into that magma, she suddenly went through the mesh, and there was an abrupt inrush of the formidable energy Sri Aurobindo called "supramental," and which Mother picturesquely described as "the boiling porridge of the supramental." Indeed, it does seem to want to pound you to the consistency of porridge.

> **59.610** — I have had a unique experience. The supramental light entered my body directly, without going through the inner or higher planes of consciousness. It was the first time. It came through the feet. . . .

A very significant detail since all yogic experiences take place above the head, at so-called higher levels of consciousness. Mother was working at the other end.

> A red and gold color, marvelous, warm, intense. And it rose and rose. And as it rose, the fever rose, too, because the body wasn't accustomed to such intensity. When all that light came into the head, I thought I was going to burst and the experience would have to be stopped. But then, very clearly, I received the indication to bring down calm, peace, to widen all this body consciousness, all these cells, so they could hold the supramental light. Suddenly, there was a split second of fainting. I found myself in another world. . . .

And this is where the confusion began (for me) because, as the experience unfolded over the years, this "other" world was no longer "other." It was our very own, the same world we see with our eyes wide open, but seen and lived differently. And after calling it "the subtle physical," Mother slipped into another terminology, then another, finally to call it "the true physical," "true matter"—the other state in matter. In fact, it was simply tomorrow's earth, much as our earth may be to an amphibian just emerging from the waters.

> Another world, but not far. It was a world almost as substantial as our physical world. There were rooms—Sri Aurobindo's room with the bed where he rests—and that's where he lived and stayed all the time; that was his home. There was even my room, with a big mirror like the one I have here, combs, and all sorts of things. And these objects were of a substance almost as dense as in the physical world, but they had their own light; they weren't translucent, transparent, or radiant, but luminous in themselves. The objects, the substance of the rooms did not have the opacity of physical objects. They didn't feel hard and dry as in the physical world....

Under the microscope, matter is not opaque at all, nor hard and dry.

> And when I woke up, I didn't have the usual sensation of coming back from far away and of having to reenter my body. No, it's as if I had been in that other world and simply took one step backwards, and I found myself back here. It took me a good half hour to realize that this world existed as much as the other one, and that I was no longer on the other side, but here, in the world of falsehood. I had forgotten everything—people, things, what I had to do—everything was gone as if it had no reality. You see, it isn't as if that world of truth had to be created from scratch; it is all ready, it's there, like a lining of our own. Everything is there. EVERYTHING is there.

Then Mother added this, which gives a sense of proportion:

> I stayed there two whole days—two days of absolute bliss. And Sri Aurobindo was constantly with me, the whole time. When I walked, he walked with me; when I sat, he sat next to me. By the end of the second day, however, I realized that I couldn't stay there any longer because the work was not progressing. The work must be done in the body; the realization must be achieved here, in this physical world, otherwise it isn't complete. So I withdrew and went back to my work.

Thus, it had taken Mother nine years after Sri Aurobindo left to find his trail. Why nine years? Because during those nine years, she went through the layers and finally reached the consciousness of the body; it was the body, the consciousness of the body that saw Sri Aurobindo's home, which no yogic or occult eyes had seen before. The eyes of the body are the ones that have access to the "other" world! For the eyes of the body, death does not exist. It is something else.

As the web grew thinner over the years and the body emerged from under its successive obstructions, intellectual, emotional, and sensory—everything evolutionary habit has put on it, i.e., the net—the "other" world was right there and the body strolled around in it "as if in the Bois de Boulogne," said Mother. Much the same way the amphibian lands on the shores of this same sunny earth, but with a different breathing mode. This is what I had so much trouble understanding, and I would ask Mother if that "other" world was not, in a way, like those spoken of by all the traditions: the Egyptians, the Greeks, the Tibetans, and the rest. But it wasn't! Perhaps simply because all these sages and mediums were in the habit of going off to heavenly "heights," or into occult depths, while the secret was located in matter—in the "feet." Clearly, no one had had the courage to go down there and stir up the vile swamp of the physical mind. Or could those sages and mediums actually have seen that same world, but *through* spiritual layers, through layers of sleep or layers of "meditation," as vague shadows of light(!) or impalpable mystical expanses, which were only the ethereal caricature, an evanescent outline of the same reality located beneath the feet? Only the body could experience "that" directly, without spiritual, occult, magic, or even electronic eyeglasses; all the world's "mystery" was just its reality approached from the wrong end or with the wrong instrument. What would a spiritually inclined fish or even an electronically equipped fish say if it looked at the earth through some aquatic meditation or through magnifying gills?

And Mother ended the description of her experience with these words:

> It would require very little, very little indeed, to pass from this world to the other, FOR THE OTHER TO BECOME THE REAL ONE. A slight triggering would suffice, or rather, a little reversal of the inner attitude. How can I say this? It's imperceptible to the ordinary consciousness; all it takes is a slight inner shift, a change in quality.

All it takes is to leave the sort of "refractive index" that blurs everything, distorts and twists and ruins everything, to enter a time without death and a space without distance, "for the other to become the real one," Mother clearly says, meaning that there is no need "to leave the world," no need "to go off." Instead, the other ray, the other non-refraction, the other vibration must *replace* our illusory and false vibration—a "substitution of vibration," she said. A slight reversal. "A slight triggering would suffice."

A universal reversal?

The human world getting out of the glass bowl?

The terrestrial fairy tale.

Years after 1959, Mother would try to explain to me that transition from one state to the other, from one "world" to the other:

> **66.263** — I don't know what to compare it with, but I'm sure there are things which, turned this way *(Mother turns her wrist one way)*, are invisible, and turned that way *(Mother turns her wrist the other way)* are visible [yes, polarized light]. It may be an internal shift, because I don't know how many times (hundreds of times), the following has happened to me: this way *(Mother turns her wrist)*, things are what we call "natural," as we're used

to seeing them, and suddenly, that way *(the other direction)*, their very nature changes. Yet nothing has happened, except something inside, something in the consciousness, a change of position. A change of position—it isn't more tangible than that; that's what is so marvelous! As a matter of fact, just the other day, I found another sentence by Sri Aurobindo: "All now is changed, yet all is still the same."[1] I read that and said to myself, "Aha! . . ." The closest explanation is that of a shift: the angle of perception is different. And it's not at all what one would be tempted to think: an interiorization and an exteriorization, not at all; it's a change in the angle of perception. You are in a certain angle, then you are in another. . . . I've seen some children's toys like that: in a certain position, they look compact, hard, and dark; then you turn them over, and they're clear, light and transparent. It's something like that.

A change in the terrestrial angle of perception?

One day, during one of the last years of her life, Mother told me:

70.29 — It's something stupendous . . . which looks like nothing.

But the experience of 1959 continued, expanded, and became increasingly natural. In 1962:

62.1210 — People are quick to say, "So-and-so is dead! . . ." I've lived that these last few days. I spent at least two hours in a world, which is the subtle physical [still that same vocabulary which was later to change], where the living and the dead are side by side without feeling the difference! It doesn't make any difference to them. There were living people, there were—what *we* call "living people" and what *we* call "dead people"; they were there together, moving together, playing together. And all that was in a lovely light, quiet, very pleasant indeed. I thought, "That's it! People have established some kind of cutoff point, and they declare, 'Now he's dead.' "

1 *Savitri*, XXIX, 719.

Seven years later:

69.175 and 215 — Something is being attempted with this body, but what? I don't know. Very strangely, it has been given a consciousness disconnected from all sense of time. There is no longer "when it did not exist," no longer "when it will cease to exist," no longer . . . Instead, everything moves together. So what is going to happen? I don't know. It's contrary to all habits. And this body is funny! Sometimes it wonders, "Am I dead or am I alive!?" This is all like a demonstration to make us grasp the secrets of existence. It's strange. For instance, I have been to places where there were a lot of people mixed together; I mean, the so-called living and the so-called dead together. They were really together, and used to being together, and finding it perfectly natural—a whole crowd of people! More and more, the impression prevails that it is our head and our way of seeing things that create strict limits, but that's not how it works! It's all mixed together.

Then this, as if the barrier were growing thinner:

69.197 — There is a place where those who have a body and those who no longer have a body mingle together, without it making any difference. They have the same reality, the same density, and the same conscious, independent existence; and there is an extraordinary similarity to material life, except that you feel people are freer in their movements. But what is strange is that I wake up and the state from "over there" continues, and it feels as real and tangible as physical things. For example, I was with someone [a person supposedly dead, right there in Mother's room], wondering, "Is this person like that physically? Is this physical?" And I was standing up! So it's as if the two worlds were . . . *(Mother slips the fingers of her right hand through those of her left hand).* Strange.

All right, the dead live. That is not really surprising. There are even a number of dead people who are more alive than many living taxpayers, while some living people are already half dead. But all the same, who are these "living people" strolling about and socializing with the "dead"? Up to now, we have hardly heard any living person tell of his *physical* wanderings with "dead people." Does it mean that, without our knowledge, a part of our being already communicates with that world (I don't know what to call it) where laws are not the same, where "death" is not the same, and which is nonetheless a physical world according to Mother's experience? Could it be that our body knows better than we do?

In any case, those who have had that kind of experience with the "dead" have generally had it during sleep, or when in certain special states, i.e., through the usual layers. But if these layers are precisely the falsehood of the world, its heaviness, its wrong or distorted angle of perception that causes all the accidents, illnesses, misery, and death of the world, what does it all mean? What is life really? And what is death really? Could there be a place in the physical, material consciousness—let's say the next terrestrial consciousness—where the nature of *both* life and death change? That would mean a really new condition on earth: neither life as we know it *nor* death as we know it.

But let us hear Mother continue her experimenting to "grasp the secrets of existence":

> **67.73** — And all that is knowledge of the consciousness of the cells.

It is not the mind, or yogic knowledge, or any form of occultism. It is a knowledge of the *body's* consciousness. The body sees. It is the body seeing its earth in a totally different way. It is the body grasping . . . its own secrets.

10

Overlife

Life and Death

Mother had countless opportunities to study death—the phenomenon of the corpse—since her early experiences in Tlemcen, described in another work.[1] One day, I asked her whether one can "experience death without dying." With her usual sense of humor, she replied:

> **68.289** — Certainly! You can even experience it materially provided the death is short enough so doctors don't have the time to pronounce you dead!

Needless to say, Mother had little regard for medical science. "In medicine, I am an atheist," she would tell me laughingly. And I am reminded of Sri Aurobindo:

> *We laugh at the savage for his faith in the medicine man; but how are the civilized less superstitious who have faith in the doctors? The savage finds that when a certain incantation is repeated, he often recovers from a certain disease; he believes. The civilized patient finds that when he doses himself according to a certain prescription, he often recovers from a certain disease; he believes. Where is the difference?*[2]

---

1. *Mother,* Vol. I, *The Divine Materialism* (New York: Institute for Evolutionary Research, 1980).
2. *Thoughts & Aphorisms,* XVII, 126.

On several occasions, Mother even had the rather painful experience of dying during a whole night, within another person, as we mentioned earlier. And then there were those countless "little deaths" when going through the web. Yet, it is precisely that *moment* of going from one state to the other that is of interest to us; that's when one gets a chance to grasp the secret—when the shift occurs. Doctors will list every possible illness, "which causes one thing or another," and cardiac arrest "which causes one thing or another . . ." but they know nothing of the reality of the phenomenon. They might as well be describing a car accident from the number of pebbles on the road. It's extraordinary how all our science is just "beside the point," a kind of mechanical caricature of "something" that totally eludes it.

Here is one of Mother's very first experiences after her first foray out of the web in 1962, as she was still going in and out of it in a constant and imperceptible back-and-forth movement, as if on the borderline between two states:

> **62.89** — It is a curious feeling, a peculiar perception of the two ways of functioning (you can't even say they are superimposed): the true functioning of the body and the functioning distorted by the individual sense of an individual body [the human fishbowl]. They are almost simultaneous, and that's what makes it so hard to explain. . . . It's as if the consciousness were pulled or pushed or placed in a certain position, and there the wrong functionings instantly appear [meaning that we're back in the web]; and they appear not as a consequence, but rather the consciousness becomes *aware* of their existence. . . .

Here, we are beginning to touch on a secret. Mother seems to be saying that the wrong functioning (which leads ultimately to death) is not the consequence of catching all the illnesses previously in the fishbowl, which had been temporarily left behind, but it results from the consciousness becoming *aware* of their existence. Illness and death are there in the fishbowl *all the time*, whether latent or manifest—that is the very definition of the mortal state—but the consciousness becomes *aware* of their existence; in other words, it gives them a reality. It is not an "illness" one contracts in the fishbowl; it is the false consciousness; that is the only real "illness" of the fishbowl. Mother continued:

> But then, if the consciousness remains long enough in that position, that has what is called consequences; the wrong functioning has consequences: little things, physical discomforts, for example; whereas, if the consciousness returns to its true position, that ceases INSTANTLY. But sometimes it's like this *(Mother runs the fingers of her right hand through those of her left hand)*: this position and that position, this one and that one, again and again, within a few seconds [in and out of the web], so I have almost a simultaneous perception of the two functionings. That's what made me understand the phenomenon, otherwise I wouldn't understand; I would only see a state of good health followed by a lapse into a state of bad health. That's not what it is. Everything, all the substance and all the vibrations must follow their normal course, you see; it's just the perception of the consciousness that is different. Which means that if we push that knowledge to the extreme, that is, if we generalize it, then life (what we generally call "life," physical life, the life of the body) and death are the SAME thing, they are SIMULTANEOUS: It's only the consciousness doing this or that, going this way or that way *(same movement of the fingers)*. I don't know if I am making sense, but it's fantastic!

It *is* fantastic. There is no such thing as "death," cancer, tuberculosis, or heart failure, but there is a false consciousness, a consciousness in a false position, which instantly *causes* cancer, tuberculosis, etc., with all their fatal consequences. If the consciousness is in the right position, it does not notice them, and there is no cancer, no tuberculosis, no illness of any kind! In other words, illness and death are constantly there—that is the normal human state—one simply is "aware" of them or one is not. All the vibrations follow their "normal course"; it is only the position of consciousness that is different. Fantastic! And Mother added this:

> And I have this experience in instances that are as concrete as possible. For example, suddenly that sort of imperceptible shift of consciousness occurs and one feels on the verge of fainting; all the blood rushes to the feet, and bing! But then, if the consciousness is recaptured IN TIME, it doesn't happen; if it isn't recaptured, it happens. I have therefore a very distinct impression that what is interpreted as death in people's ordinary perception, all the external signs and everything, only indicate a failure to restore the consciousness to its true position fast enough. . . . I fully realize that words are totally inadequate to express the experience, but this is perhaps on the way to knowing the "thing" [death]; knowledge implies a power to change things, doesn't it? And I do feel that something is leading me toward the discovery of that power, that knowledge, naturally through the only possible means: the actual experience. And with a great many precautions, because I can sense that . . .

It is dangerous, to be sure. One may not recapture the right position fast enough. But the essential fact is that "life and death are the same thing." It has nothing to do with cancer or ninety years of wear and tear "which causes this and that." Which means that all medical science is wrong! We are stuck in a mortal glass bowl—the doctors are 100% right—but they are merely treating an illusion.

Now the whole question is to understand that change of position.

Mother's comprehension of the process was further enhanced by a strange experience triggered by the death of a disciple. Briefly, a disciple was walking in a state of inner concentration, paying scant attention to the material world. He bumped into something, tripped, and fell, fracturing his skull. After attempting some horrible operations, the doctors pronounced him "dead." While all this was going on, the disciple came to Mother in a very alive consciousness: He was near her, peaceful, as if pursuing his meditation. But suddenly, Mother felt a violent shudder in the disciple and he disappeared, at the exact moment his body was being burned on the funeral pyre. And Mother exclaimed:

> **62.47** — In the state he was in, it made NO DIFFERENCE to him whether he was dead or alive; that's what is so interesting! And it is because they burned him that he was violently put in contact with the destruction of the form of his body. . . .

One could say he suddenly became "aware" that he was dead.

> *(Question:) In terms of your experience, what conclusions can be drawn from this story?*
>
> Well, it means you can die without knowing you're dead! He went on being, living, experiencing, absolutely INDEPENDENT of his body, without needing his body in the least to pursue his experience. I find that an important experience. . . . So one could say it is necessary to die unto death to be born to immortality. Dying unto death means to become incapable of dying, because death no longer has any reality.

The position of consciousness changes, and not only do cancer, heart failure, and the rest no longer have any reality—meaning they cannot be, or manifest themselves, even though they're still there, latent in the fishbowl—but death cannot be either. Death is still there, but dying or not depends upon a position of consciousness, and the same holds true for accidents and all the rest.

Then this experience of the "death of death" became clearer:

63.163 — The impression in ordinary life (few people are aware of it) is one of being "subjected" to a destiny, to a fate, to a will, to a set of circumstances—words make no difference; it's something weighing on you and seeking to manifest through you. But since that experience of the "death of death," I have the impression . . . Before, when I worked on people, either to keep them from dying or to help them once they had died (hundreds and hundreds of things I was doing all the time), I did it with the sense that death was something to be overcome or subdued, or something whose consequences one had to correct. . . .

One overcomes or subdues an enemy, thereby giving the enemy a great deal of strength by struggling with him, but what if there is no enemy! What if there is nothing except an illusion?

Now my position has changed. But it may sometimes take years to translate into a conscious power. In this case, the conscious power would be a capacity equally to cause death or to prevent it, to set in motion the necessary forces—ALMOST A MECHANICAL ACTION ON THE CELLS. This is what that power would mean: you can cause death; you can prevent death. And there is no longer the usual sensation of a sharp opposition between life and death, its opposite; death is not the opposite of life! I understood that then, and I have never forgotten it; death is NOT the opposite of life. It is like a change in the cells' functioning or in their organization. And once you've understood that, it is very simple: you can easily keep it from going this way or that *(Mother runs the fingers of her right hand through those of her left hand, on one side of the web, and on the other)*; you can do this, or that. Evidently, this would mean a new phase of life on earth.

62.117 — This "dying unto death" was so clear; it was overwhelming with power! And also the impression: easy, it's easy! It isn't a question of easy or difficult—it's spontaneous, NATURAL, and so smiling!

That's just it—natural. It is the natural state par excellence. We have entered a sealed bowl of irreality where we become aware of all kinds of disasters, which naturally happen as a result of our being aware of them, just as my death would have happened in that canyon had I become aware of it, or had my body believed it was going to be killed. But strangely, at that moment, it was as if there were nothing, and so there *was* nothing! There was no accident. For a minute, I was in the natural state. For death to occur, there has to be a contact with death, but what if there is no contact?

"A change in the cells' functioning . . . almost a mechanical action on the cells . . . a wrong position of consciousness not corrected fast enough"—we keep coming back to that crossing through the web of the physical mind. Years later, Mother was coming closer to the key:

> **66.262** — The problem for me is to unravel the process in order to have the power to undo what has been done [death, this whole enveloping web of irreality]. After all these years, there is something in me that wants to have that power or that key: how it works. And it may be necessary to feel or LIVE how things turn out this way *(Mother turns her wrist in one direction)* in order to be able to do that *(she turns it in the other direction)*. What is remarkable is that now that this mind of the cells has become organized, it seems to be going back, very fast, over the entire course of human mental development in order, precisely, to arrive at . . . the key.

It is the mind of the cells that holds the key to death, or rather to non-death, to the state in which death *and* life change into something else, in which the opposition no longer exists.

Death is not the opposite of life! It is the same state, the same brew of "something" we call existence, where we sometimes catch death for good. But, in fact, it is always there; we are born with it; we are born *in* it, one could say. Our cells forever spin the habit of defeat and death—their "stupid goodwill." But if we change that vibration, that spinning mode, and give them instead another, sunny, free vibration to repeat, then everything changes! Then there is no longer life as we know it, which is only suspended death, with false time, false space, false matter; and there is no longer death as we know it, which is only the disappearance of our false way of seeing and false material stage, but something uninterrupted, with or without a body, in a true time, a true space, a true material and earthly matter. That is "overlife," the fishbowl in pieces, which does not mean the death of the fish but the beginning of another species or another kingdom on earth. Yes, a "new phase of life on earth."

> **70.31** — What I have learned: The religions have failed because they were divided; they each wanted you to be religious while excluding all the other religions. Likewise, all knowledge has failed because it was exclusive. And man has failed because he was exclusive. What the new consciousness wants is no more divisions. To be able to understand the ultimate spiritual, the ultimate material, and to find the meeting point where . . . they become an effective power. And that's what the body is also being taught, in the most radical ways. They all say, "This, but not that." No! This AND that, and also this other and that other and everything together; to be flexible enough and wide enough to encompass everything, including in the body. The body is accustomed to: "This, but not that; this *or* that" No, no, no—this AND that. That's the great Division, you know: life versus death. And everything stems from that. Well, words are woefully inadequate, but overlife is life and death *together*. Why even call it "over*life*"? We're forever tempted to lean on one side: the light versus the dark ("dark," well . . .).

Suddenly, I am reminded of this strange verse by the Vedic Rishis, going back five or seven thousand years: "He uncovered the two [worlds], eternal and *in one nest*" (Rig-Veda, 1.62.7).

The whole question now lies in this "with or without a body," that is, whether the body will have the power or capacity to cross over to the other state and gradually transform its old conditions into a new condition, whether the body can remain a living link between the two worlds: to live where the living and the dead are together, "without it making any difference," or whether it will continue its old habit of disintegration, break its shell and "die" so the human being can return again and again into the web, until he finds the key to the illusion. Why the illusion? So that we may find what no contented species could find before us (probably because it was too contented within its species): the power to undo the genetic weave that ties us to a certain species and locks us into a single way of being, while the goal of evolution, if there is one, is to be everything and live everything and integrate all the ways of being, those known and those still unknown, within a single, free, and happy individual being, who has no shell but is yet material.

This power is the mind of the cells.

A Dangerous Unknown

A strange, and painful, life was about to begin for Mother. It is very easy to speak of "the next species" and to put it all into paragraphs (or is even that so easy?), but on a day-to-day basis it is very agonizing for the pioneer. Is one even going somewhere? Is this insanity, disintegration, or something else? There is no one to tell you. Perhaps her only human consolation was to be able to talk to me, but soon they would even close her door to me. Another species is truly crazy. Honestly, I know of no one more heroic than Mother.

And yet she laughed and made fun of it all. How she made fun of it!

> **70.294** — The body says, "Actually, it's mostly for others that it would make a difference [if Mother died]! For me . . ." They, you see, are still living in this sort of illusion of death as a result of the body disappearing; but even the body doesn't exactly know whether this is true anymore! In its view, matter should be the truth, but it isn't too, too sure of what it is! There is the other way of being. It knows that the old way isn't right anymore, but it is beginning to wonder exactly what the other way will be like. What will be the relationship of the new consciousness with the old consciousness of those who remain human? It comes—it's curious—it comes like a breath and disappears. The body is in pain . . . a funny kind of pain; it groans, literally, as if it were in terrible pain, then a little something happens and the pain disappears, to be replaced by something that is not what we call bliss—I don't know what it is—it's something else, but extraordinary: something new, completely new. And all that takes place in some kind of netherland, which is no longer one thing, but not yet the other. This is no longer the body consciousness as it usually is—oh! it is on the way to something else, but it isn't there yet. But the presence of the Grace is absolutely marvelous, because if I were not given the true meaning of what is happening as it is happening, the experience in itself would be constant agony. It is the old way that is dying.

Eight years earlier, she had told me this:

> **62.126** — It has reached the point that if I didn't care for people's peace of mind, I would say "I don't know whether I am alive or dead!" Because there is a life, a type of life vibration that is completely independent of . . . [Mother was about to say "the body"]. No, let me put it another way: The way people usually feel life—feel they're alive—is intimately connected with a cer-

tain sensation they have of themselves, a sensation of their body and of themselves. Now suppose you completely remove that sensation, that type of sensation, the type of relation that people call "I'm alive"—you remove that—then how can you say, "I'm alive" or "I'm not alive"? That no longer exists! I can't say, "I'm alive" as they do; it's something else entirely. . . .

The "agony" was to be a long one.
And Mother added, laughingly:

You'd better not keep a record of this conversation, because they [the disciples] might wonder if I should not be treated as a mental case! But that doesn't matter, either! . . . What I am saying is becoming more and more difficult. Perhaps in fifty years people will understand?

Clearly, Mother was no longer living in our ordinary "I'm alive" world, but where was she, then? In "death"? And what is death, actually? One day, I asked her the question and received an answer that rather stunned me, although I had long been prepared by what she had already said: "Death is not the opposite of life."

67.73 — I have come to the conclusion that there isn't really such a thing as death. There is only an appearance, an appearance based on limited vision. But there is no radical change in the vibration of consciousness. The importance given to the difference of condition is only a superficial difference, based on ignorance of the actual phenomenon. A person capable of keeping some means of communication could say that, for himself, it does not make much difference. But all that is still in the process of being worked out. Some areas are still unclear and certain experimental details are still missing.

(Question:) You say there is no difference. Does it mean that one still perceives this physical world from the other side?

Yes! Yes, exactly.

One perceives people and . . . [I wanted to say trees and birds in the sky and the beautiful sun over our earth]?

Yes, exactly. Only, instead of having a perception . . . one sheds a sort of illusory state and a perception that perceives only appearances, but one does have a perception. There were times when I had the perception; I was able to see the difference, but the experience wasn't total, you see; not total in the sense that I was interrupted by external circumstances. But there is perception. Not exactly identical, though SOMETIMES WITH A GREATER INTRINSIC EFFECTIVENESS. But this isn't really perceived from the other side. . . .

And Mother added this, which completely opens our eyes:

But all this is clear, precise, and EVIDENT, only with the new cellular vision, because—how can I put it?—I knew all that before [Mother had had many so-called occult experiences], but now I have seen it again in the new consciousness, with the new way of seeing; and the understanding was total, the perception was total, and totally concrete, with convincing elements that were completely lacking in the occult knowledge. This is knowledge of the cellular consciousness.

It is the body, the body's consciousness, that connects directly with the other side of the fishbowl. Why, of course! This is not soaring into pure spirit; this is going into matter itself, as it really is. The dead are there. Death is here, with us. And it isn't "death" at all!

Then, another day, Mother made a rather enigmatic remark, which is quite illuminating when we think about it:

70.253 — To the body consciousness that remains conscious when the body is asleep [and what is that consciousness, if not the consciousness of the cells?], the world as we know it is always dark and muddy. In other words, it's always gloom—you can hardly see—and mud. And this is not an opinion, not a sensation; it is a material fact. Which means that [cellular] consciousness is already conscious of a world that would no longer be subject to the same laws.

72.267 — When I remain immobile, like this, after some time, I feel a whole world of things being done, being organized, but it's—how can I say it?—it's another kind of reality, a more concrete reality. How is it more concrete? I don't know. Matter seems like something unsubstantial next to that [but then how would the fish's water seem next to a sunny meadow?]. Unsubstantial, opaque, unreceptive. Whereas that... The funniest part is that people think I am sleeping! I am almost no longer a part of the old world, so the old world says, "She's finished!" I don't care at all!

One might think that, gradually, Mother was going over to the side of the dead, as if this whole evolutionary journey, all these efforts, all the pain of the earth over the ages, were to end up in a state that, although material, had no link, no connection or continuity with the material evolution of the species. But that is not the case. Mother was not going off into "death." It would appear that, at the cellular level, a curious blend of alchemy takes place that alters not only life as we know it, but death as well. Truly another state in matter. The dead have no cells. And if the little animal cell has toiled over three and a half billion years on earth, it is not to end up vanishing into thin air; the cell, too, must fulfill its evolution. It may even be where the next world is being built, a world that has nothing to do with our type of life, or death.

72.127 — I feel I am becoming a different person. No, not just that: I am touching upon a different world, a different way of being, which might be called a dangerous way of being. Dangerous, but marvelous. The feeling that the relation between what we call "life" and what we call "death" is changing completely, becoming drastically different. You see, it isn't that death is disappearing (death as we conceive of it, as we know it, in relation to life as we know it); it's not that, not that at all! BOTH are changing into something yet unknown, which appears both dangerous and absolutely marvelous. We tend to want certain things to be true (what WE consider favorable), and others to disappear, but that's not it! It's EVERYTHING becoming different. Different. From time to time, for a brief moment, there's wonderment, and then, immediately, the sense of . . . a dangerous unknown. That's how it is. I spend all my time like that.

72.912 — Everything is crumbling, except for—for what? The divine . . . something—what? It's like an attempt to make one feel there is no difference between life and death. That's it. There is neither life nor death, neither what we call life nor what we call death. There's . . . something. And that something is divine. Or, rather, it is our next step to the divine.

69.164 — It's strange, how it all seems to be the same, yet it's becoming very different.

62.132 — For those people coming in one or two hundred years, it 'll be easy. They will only have to choose: to belong to the new system or the old one. But now! A stomach has to digest, you see. . . . Is this all folly? Or is it something possible? I don't know. Nobody has done it before, so there is no one to tell me.

70.44 — The body feels . . . the word "anguish" is too strong, but it's the impression of having reached the point of . . . the unknown, the unknown, the . . . something. That's a very, very strange sensation. It's like a new type of vibration, one could say. It is so new one can't call it "anguish," but it's the unknown. The mystery of the unknown. And this is becoming constant. So the only solution for the body is total abandon; and it is in this state of total abandon that it realizes that this vibration is not a vibration of disintegration but something—what? The unknown, totally unknown—something new, unknown. Sometimes, it is seized with panic. And it's not that it suffers much; I can't call it suffering; it's something . . . truly extraordinary.

Indeed, the "other thing" must be so much "other" that it must feel like death to the body!

It is the equivalent! It's a funny kind of life, at any rate. I will soon be dangerously contagious, you know!

It may well be that the world itself was going to catch the dangerous contagion.

70.114 — It's a very strange impression, as if one were on the verge—but the verge of what? I don't know. Something . . .

70.277 — Something that has an innumerable experience all at once.

72.221 — At times, the body feels this is impossible—one cannot exist like this—but just at the last minute, something comes and there's . . . a harmony absolutely unknown to this physical world. A harmony—the physical world looks horrifying in comparison. It's really like a new world trying to manifest itself.

72.135 — Never have I had such an impression of . . . nothingness—the nothing. Nothing. I am nothing. It is as if the body could die at every minute, and at every minute it is miraculously saved. That is truly extraordinary. Along with a constant perception of world events, as if everything were . . . *(Mother puts the fingers of her right hand tightly between those of her left hand)*, as if there were a connection.

73.173 — I sometimes wonder how it is possible. . . . Sometimes it is so new and unexpected, it's almost painful.

71.258 — It is like standing on a ridge, and the least misstep would send you tumbling into the abyss. Everything seems different. The nature of the relation with others is different, the nature of everything is different, but what, what? It's like a balancing act: an incredible power and, at the same time, an incredible impotence; you know, like being suspended between the most wonderful and the most vile. Like that. I don't even know where I am going—whether I am going toward transformation or toward the end.

70.31 — The body feels so distinctly that it does not belong here anymore, yet it isn't there either, so . . . In its appearance, this body is something totally absurd, with apparent weaknesses that human beings despise and unbelievable strengths that human beings cannot withstand.

Now we come to the real problem. A new species has to be bearable to the old species. Can one, alone, change to another species?

71.37 — It's as if the two extremes—a marvelous state and a general disintegration—were together, intermingled. Absolutely everything is disintegrating: people you count on let you down; dishonesty seems to be spreading everywhere. And at the same time, for a flash, comes a wonderful, unimaginable state, as the extreme opposite. As if this was trying to take the place of the other, but the other fights back ferociously. And all the circumstances are like that, all the people are like that, from the government down to the people here. And then, for a few minutes, that wonderful state comes into my body, then goes away. This is the situation. This is what I live night and day, all the time. Three minutes of splendor for twelve hours of misery. In other words, the extent to which the world is not what it should be is becoming very,very acute. Usually we say: "There are both good and bad things in the world," but this is all childishness; the good things are no better than the bad ones! That's not IT. The divine is something else.

The "divine" is the next way of being on earth.

A dangerous unknown, the very battle of the world—the one being fought in a hundred countries, under a thousand banners, a thousand pretexts and slogans, but it is the battle of the next species on earth. Will this earth accept the future or will it be engulfed, one more time, in a cataclysm only to begin again, here or elsewhere, the eternal evolutionary quest for love in freedom and joy?

Indeed, when there is no longer life *and* death, a formidable wall will fall away from our consciousness—like the walls of Jericho—and with it, the most ancient anguish on earth.

11

MOTHER'S DEPARTURE

Why did she leave? Why?
For so many years now I have come up against that pain.

70.295 — For the ordinary consciousness, this appearance *(Mother points to her body)* seems to be the most important. Obviously, it will be the last thing to change. But to the ordinary consciousness, it seems to be the last to change *because* it is the most important: It will be an unmistakable sign. But it's not like that at all! It is the change in the CONSCIOUSNESS of the cells that is the important thing. All the rest is consequence. For us, when this *(the body)* is visibly something other than what it is now, we'll say: "Ah! Now the thing is done." That is not so; the thing IS done. The body is a secondary consequence.

It is clear that once the mind of the cells has begun churning "that other vibration," the mantra, it will repeat it and churn it as imperturbably as the cells have churned out amino acids for three and a half billion years, or as the hydrogen nucleus has churned its lone electron—provided the present cells last long enough to implement the transformations that naturally result from the new vibration. "Give me time" was the prayer so often repeated by Mother. "Give me time." "One wishes one had hundreds and hundreds of years to do the work!" she said a year after the passing of Sri Aurobindo.

60.281 — I am prepared to struggle for two hundred years, but the work will be done.

Yet even that kind of "time" did not seem really a problem.

54.258 — The cells that can vibrate at the touch of divine Joy are regenerated cells on the way to becoming immortal.

67.2110 — I have a feeling that death is only an old habit now; it is no longer a necessity. It's just that the body is still unconscious enough to feel the need for total rest, that is to say, inertia. When that is abolished, there is no disorder that cannot be remedied, or at least no wear and tear, no deterioration, no disharmony that cannot be remedied. That's the only problem. And then that enormous collective suggestion weighing on me.

This suggestion, on the one hand, and the old memory of the peace of the mineral, on the other. Though even that memory was abolished and replaced by the immobility of the "lightning-fast waves," so fast they seem immobile.

61.206 — *(Question:) When everything is still, like now, and nothing seems to happen, is something happening?*

Something happening? I don't know. This is the same infinity as when leaving one's body. And that, in itself, *is* something. It's very difficult for the body to have it; there's always something quavering or stirring. It's like putting everything back in place, yet nothing moves. And it isn't just silence; it's IMMOBILITY WITHOUT TENSION, without effort, without anything. It's like eternity in the body. It is a state that is perfectly natural to me; I can hear the clock chiming.

64.189 — The years and months are going by at dizzying speed, without leaving a trace; that's what is interesting. So if you observe that, you begin to understand how it is possible to live almost indefinitely, when the friction of time is gone.

And again in 1970:

70.1410 — The consciousness of the body is slowly changing, to a point where all of its prior life seems alien; it seems like someone else's consciousness, someone else's life. It's like having no past, you know—totally focused ahead with nothing behind. A curious sensation. A curious sensation that something is beginning, not in the least the feeling of something ending—something beginning. With all the unknown, all the unexpected. . . . Strange. I have this constant feeling that things are new, that my relationship to them is new.

67.1551 — Not a day passes by without the verification that, not even a dose, but just a tiny dose, an infinitesimal drop of "that" can cure you in a minute. For instance, a wave of disorder breaks out; the substance constituting the body begins to feel it, then to be affected, then everything begins to get disrupted. It is this disruption that prevents the cohesion of the cells necessary to form an individual body. Then you know: "This is the end." So the cells aspire, and there's a kind of—well, it's as if the wave of disorder became denser and something stops: first, a feeling of Joy, then a light, then harmony, and the disorder is gone. And immediately a feeling comes into the cells of living in eternity, for eternity. Now, that happens not only every day, but several times a day. That's what the work is about. It is a very obscure work. Actually, all the proclamations, the revelations, the prophecies are very comforting, you know; they give an impression of something "concrete," while this is very obscure, invisible (it won't have visible results for a long, long time), not understood. In fact, to the extent that it is really new, it is not understandable.

It seemed obvious to me that the experience would go to the end; there was not the slightest doubt in my mind. It was both simple and obvious. I even visualized the next logical step in the process to be the discontinuation of all food intake (Mother was taking only liquids), and the abolition of the old digestive system, to be replaced by direct absorption of energy. But, in fact, I had not understood the real problem. I still believed in some "marvelous and miraculous transformation," the much-awaited, tangible proof, visible to all, which would compel this recalcitrant humanity to understand the whole process, to understand that there *is* an issue, a logical and rational way of leaving this suffocating fishbowl and creating a new life on earth. It was not just Mother's body I was thinking of, it was the earth's body. This pain-ridden, miserable, little body had to understand, at last, its own freedom and joy, and *the* way out.

I had not even understood that the first earth and the first humanity were right there, around Mother, in the form of a number of disciples who in fact represented the earth, who *were* the human samples for the great evolutionary operation, and if it did not happen *there,* then where would it happen? If it were not understood *there,* then who would understand? I was still visualizing an "abstract" earth.

And I had not understood a second aspect of the problem, the essential one, in fact—the one that *could* change humanity and compel the earth in spite of itself to shift into accelerated evolution, as it were, and reach a point, a moment, an inevitable juncture when all the old chaos would collapse onto itself like a dying star and the new door would open.

That aspect is "power."

We can see the acceleration everywhere around us.

But acceleration is very painful, grating. Everything grates.

And power is something quite unbearable.

Mother was becoming "unbearable" to all the little evolutionary samples gathered around her. One does not open one body to the fantastic energy from which we are so cozily protected inside our web without that energy radiating and spreading "contagiously" to all surrounding matter. I knew myself, because I experienced it in my own body every time I was with Mother (and even from afar). Being near her was literally like being immersed in a bath of lightning, a flood of power so compact and so dense, one's whole body seemed on the verge of melting. One was seized from within, in each cell, as if thousands of years of darkness and pain suddenly were calling out and praying for light, for love, for vastness, for freedom; and one plunged body and soul into that bath of fire, as if one had finally come in touch with "that" with one's body, "that" which one had longed so much for, aspired so much to, through life after life of pain and despair, through millennia of pointless futilities. And there one was at last. But one had to *take the plunge,* to *melt,* because if one didn't, if there was any resistance, the least form of "me" amid that flood of power, things would break or grate or revolt. It became unbearable.

And all the little samples around her were grating in chorus. And all the earth around her was fighting and kicking.

67.34 — When that luminous power comes in, it is so compact! So compact, it gives the impression of being much heavier than matter. It is veiled, veiled, veiled, otherwise—unbearable.

68.1311 — There's only one thing: like an accumulation of force . . . a force that could be a power. I feel it is slowly, gradually accumulating, along with a very clear awareness of every obstacle, every opposition, the general human attitude. And a distinct perception that one must remain veiled. This is a time to remain veiled. That's all.

70.165 — If there were a certainty; if, for example, Sri Aurobindo told me, "This is how it is," then it would be very easy. But what's hard is, I am surrounded by people who believe I am sick, you see, and who treat me like a patient, when I know I am not sick. I am surrounded by a certainty that I am very fast approaching the end, so this poor body is a bit wobbly.

71.177 — When things quiet down and I can return to my normal atmosphere, it's as if everything dissolved; the pain stops. But it rushes back from the outside with a vengeance: people quarreling, circumstances going astray, everything. And all that is dumped on me, so . . . It's an onslaught of falsehood.

71.63 — "She's old, she's old"—that creates an atmosphere of resistance to the change. It almost creates a conflict within the being. "It's impossible, it's impossible," from every side.

71.33 — You know what I think? They're all old and I am the only one who's young! As long as they are what they call "comfortable," with the freedom to do stupid things they wouldn't do in the outside world, they are satisfied. And yet, there is the feeling one COULD hasten the advent, if one were a conqueror. Basically, they don't care.

69.511 — I no longer have any control; they have each seized control. I have lost the habit of saying, "I want."

66.179 — I feel I am hanging by a delicate thread amid an absolutely rotten atmosphere riddled with disbelief, futility, ill will. It's exactly that, a thin thread, and it's a miracle that . . . They don't even understand that if that vibration of truth were to impose itself, it would be the end of them! The marvel is this infinite compassion that will not destroy anything. It simply waits. It is there with its full power, its full force, and it simply makes its presence known without imposing it, to keep the damage to a minimum. It's a marvelous compassion. And all these idiots call it impotence!

65.1610 — They wear a mask of goodwill, but the inner vibrations still belong to the world of falsehood.

64.221 — It's an absolute farce, you know! And it has been going on since 1926. There must be—I am being very generous, patient, and lenient—a good one third [of the Ashram population] who are here solely because they're comfortable. They work if they want to, they don't work if they don't; they're always fed, always sheltered, clothed, and basically do pretty much what they please (all they have to do is feign obedience). But the moment they are denied some convenience, they start grumbling. Yoga is out of the question; it is a million miles away from their consciousness; they talk of nothing else, but it's only talk. When I say "no" to something, they pretend they've heard "yes," but anyway . . . That's "spiritual" life for you!

64.3010 — The people around me are not helping. The people nearer to me have no faith.

61.254 — I am *not* a group leader. Lord no, not for the world! It's disgusting. I want to make a statement: "I am not a group leader. I am not the head of an ashram!" Sometimes, I feel like saying shocking things; how well I understand Sri Aurobindo, who went to the other side!

62.132 — It's people's thoughts that are a nuisance! All these people, all of them, constantly thinking: old age and death, death and old age and illness, oh! . . .

But I did not really fathom the extent, the depth of the negation:

69.105 — There are minutes when the body feels it has escaped the law of death, but it doesn't last. People come in with all their thoughts, and because of that, it's a bit hard. There's a considerable number of desires that it die, you know! Everywhere, they are everywhere! And the body sees that; it sees it. I am not entirely sure that all the aches and pains it has everywhere, constantly, do not come from . . . are not the result of all these ill wills.

68.155 — I've struggled and struggled, but there are too many lies around me.

And then this cry:

69.234 — It's the whole system that should be dissolved!

And in 1972:

72.103 — The atmosphere is dislocated. We come claiming to preach unity to the world, so at least out of decency we should give the example! We give the example of everything that should not be done. I see, I really see; if I were to leave, I have no one here; it would be our destruction.

In truth, it was neither biology nor physics nor Mother's ninety-five years; it was simply that she could no longer stay.
Nor could Sri Aurobindo:

65.412 — It is his compassion that made him accept the people around him as they were, otherwise it was a great suffering for him.

Sometimes, it was heart-rending:

68.156 — I look at this body, and sometimes, when there is too much incomprehension, when people around me are really too lacking in understanding, it says: "Oh, let me go! All right, never mind, just let me go." Not tired or disgusted, but . . . And it's really pitiful. So I say to it, "No, no, no!" as one talks to a child. It is a question of patience, you see. What is going to happen? I don't know. In any case, you, for one, will know. You will be able to tell them: "It's not as you think." I would tell them myself, but they won't listen to me. I don't know, I don't know what's going to happen. What is going to happen? Do you know?

Some day it will be glorious.

When you're doing something for the first time, no one can explain it to you.

A question of patience.
They had no more patience. They were even grumbling all around her.
The whole earth was grumbling.
"I have no one here."
This was Sri Aurobindo's ashram.
And then, one day, they closed Mother's door on me. She

had no one left to communicate with.

> **69.245** — You are the only one whom I can speak to. The others don't understand.

She was alone with her "guardians."
That day they sealed destiny.

Mother had foreseen the world's resistance. She had also seen the need for a long period of immobility, "in the undulation," without the constant intrusion of ill wills from the outside.

> **72.262** — I think the body is overly sensitive now, and it needs to be protected from all the things coming from the outside—as if it had to work from within, like inside an egg.

That was in 1972, a year before her departure.

> **69.2412** — If someone comes in my room who is upset with something I have said or done, all the nerves suddenly feel under torture. And it comes from the person who is there, who presents all outer signs of devotion, etc.; there's absolutely no outer sign of anything, no spoken or direct manifestation of any kind, but all the nerves feel under torture.

Five years earlier, in 1967, in the middle of a conversation, Mother had suddenly stopped, asked me to take a paper and a pencil, and began to dictate in a perfectly neutral voice, as if she were talking from "elsewhere":

> **67.141** — Because of the necessities of the transformation, it is possible that this body may enter a state of trance that appears cataleptic. Above all, do not call doctors! Nor should anyone rush to announce my death and give the government the right to intervene. Keep me carefully sheltered from any deterioration

> that might come from outside—infection, poisoning, etc.—and have tireless patience. It may last days, perhaps weeks, and maybe even longer. You will have to wait patiently for me to come out of that state naturally, after the work of transformation is completed.

Cataleptic trance, meaning total immobilization, with cessation of the heartbeat and all the appearances of death—something every yogi knows.

On her orders, this note was distributed to five people among those closest to her. Hence they knew. Indeed, Mother had seen the whole picture.

> **65.412** — According to external science, one burns toxins during sleep; similarly, this immobility illuminates the obscure vibrations.

A whole "cryogenic" science has even developed in recent years as a cure by cooling the tissues; a cataleptic trance is the same process, only natural.

Then, in April 1973, just one month before the disciples closed Mother's door on me (oh, I was so, so unaware of the jealousies around me; I lived near Mother without noticing anything, in that wonderful tale of the future, thinking that everyone understood; it was so obvious!), Mother suddenly made the following remark:

> **73.74** — I seem to be gathering all the resistances of the world. You see, I have a solution for transforming the body, but it has never been done before, and it's so . . . incredible. I can't believe this can be it. Yet it's the only solution for me. The body feels like going to sleep ("going to sleep" in a special way: I remain fully conscious), and awakening only after it is transformed. But people will never have the necessary patience to go through that, to take care of the body. Everyone will think it's the end and they will stop taking care of me.

But of course, Sleeping Beauty! It was obvious; it made perfect sense. Mother was therefore preparing to go into that cataleptic trance. Yet two months earlier, in January 1973, Mother had had a vision which she had described to me breathlessly: She was being buried alive. It was the third time she had had that vision.

> **73.101 and 72.54 and 69.245** — Oh, I haven't told you! It was yesterday or the day before, I don't remember. Suddenly, for two or three minutes, my body was filled with horror: the idea of being put, like this, in a tomb was so horrible! Horrible. I couldn't have stood it for more than a few minutes. It was horrible. And it wasn't just because I was buried alive, but my body was conscious. It was supposedly "dead," according to everyone, because the heart had stopped beating, yet it was conscious. That was a horrible experience. . . . I gave all the signs of death, that is, the heart had stopped, everything had stopped, yet I was conscious. They should be warned not to . . . not to rush to . . .

Then a second time:

> Because it may be . . . temporary, you understand? It may be momentary. You understand? Do you understand what I mean? I feel an effort is underway to transform this body; it feels it. It is willing, but I don't know if it will be able to do it. You understand? So it may give, for a while, the impression that it's over, but it would only be temporary. It could start up again. Because I may not be able to speak at that moment and to say this. So I am saying it to you. I don't know; I don't know what will happen! It is so difficult, at times, that I wonder if the body will be able to withstand it. But I wish someone could be there to prevent that blunder, because all the work would be lost. There must be someone with authority who could come forward and say, "You mustn't. Mother doesn't want this." You?
>
> *But who will listen to me? They'll say I'm crazy! They won't even let me come into your room!*

I did not know how prophetic my words were. On May 19, 1973, the door closed on Mother. She was alone. I was alone. She had six more months to go. Soon, I was going to confront the whole pack. There was all of Mother' Agenda, so dangerous for the "disciples," a secret about a future that had nothing to do with their spirituality. I have been slandered, followed all the way to the Himalayas, threatened with court cases, denounced to the government of India, and harassed by the police. And I don't know who sent those men to kill me in the canyons. That's "spiritual" life for you, as Mother would say.

They have even printed a false "Agenda" to prevent the publication of the authentic one.

The old anthropoids are quite merciless toward those who do not conform to their tribe.

But even with the door closed, I could not believe it was over. Those cells could not die. The world could not throw such a wonderful hope into a grave!

> **73.283** — And the material consciousness repeats: OM NAMO BHAGAVATEH. . . . It's like the backdrop of everything: OM NAMO BHAGAVATEH. . . . You know, a backdrop that is a material support: OM NAMO BHAGAVATEH . . .

No, those cells could not die.

> **69.245** — Leaving isn't a solution! I wish . . . I wish they would not put me into a box and throw me into . . . just like that. Because even after the doctors have pronounced it dead, it will be conscious; the cells are conscious.

Then, the morning of November 18, 1973, someone came to inform me that Mother had "died" in the night, that she was lying in the ashram downstairs hall, and everyone was filing past her.

I arrived there stupefied. She was lying beneath golden neon lights whose heat was reflected off the zinc ceilings, while fans whirled amid the deafening murmur of the crowd. They had brought her down there barely seven hours after her "death," removing her from the peace of her room and from her atmosphere, and tossing her like fodder to those thousands of vibrations of anguish, of grief, of fear—of falsehood.

Three ashram doctors had pronounced her dead. It was medical and irrefutable.

A few days earlier, on November 14, around midnight, she was on her chaise longue—she was so bent, she could no longer lie on a bed—and she had asked to walk: "I want to walk, otherwise I'll be paralyzed." She walked, supported by one of the guardians, until she turned blue. During the night of November 16, she asked again to walk: "I want to walk."

Those were her last words.

I want to walk.

But in that tomb where they have put her, I know of cells that are repeating: OM NAMO BHAGAVATEH . . . OM NAMO BHAGAVATEH . . . OM NAMO BHAGAVATEH . . .

And they will repeat their invocation over and over, until the earth comes out of its unreal falsehood.

Until it comes out of its false materialism as well as its false spiritualism and emerges into true matter and the divine life on earth.

Though perhaps some surprises are still in store for us.

As early as 1958, she had said, "Wait for the last act."

12

APOCALYPSE OR FAIRY TALE?

What is going to happen?

We all know the world situation: The population of China has just reached the billion mark—one thousand million people. Every year, India makes twelve million more babies. It is a geometric progression. No human device can stem this tide. I have seen entire mountain ranges in the Himalayas stripped clean of their trees, within twenty years' time. It makes you shudder. Attila was a joke in comparison. The whole earth is full of little Attilas; indeed, one wonders if these are humans, or something else disguised in a human body.

Perhaps this is the real problem: The earth is full of beings that are not human; they are goats or maybe rats or rabbits, but not humans. They may have science, democracy, and religion, but they are not human. They are just very ingenious digestive tracts. No species is more false. A rat is what it is, without any pretense. Man is not what he is; he pretends to be a lot of things, with a Bible in his hand and a tie around his neck. Man and falsehood go hand in hand.

Or rather, we are *not yet* men.

Our falsehood is in the process of exploding in our face. That is the only phenomenon happening in the world. Man is in the process of becoming what he is, and what *is not* will simply be removed from reality. How?

That they will be removed is beyond a doubt.

But then there are millions upon millions of liars, and falsehood is so tightly interwoven with truth that one can hardly imagine how it would be divinely possible (with emphasis on "divinely," because humanly . . .) to untangle this mess without destroying the good along with the bad. Though, on closer look, one realizes with Mother that "the good is not any better than the bad"; it is the same quagmire of "something" which is not what it should be, either at its best or at its worst. In evolutionary terms, it is a certain cellular combination—neither good nor bad—that has garbed itself in intellect, philosophy, microscopes, religion, and a number of other ingredients that we can judge as much as we like, but our judgment of them has no essential relevance to the species itself, anymore than the gospels or misdeeds of little fishes had any relevance to the making of mammals. So as far as the "sorting out of the Just" is concerned . . . well, where exactly are those Just? The Apocalypse?

> **60.237**—There are even people who foresee the end of the earth, but that's nonsense [said Mother in her simple language], because the earth has been created for a certain purpose, and it won't disappear before things are accomplished. But there might be . . . changes.

What does the little cell "think" of all this? Perhaps that is the only true and relevant question. The cell may even be the site where we will uncover what man *is*, without device and falsehood, and especially—thank God!—without "truth." Clear life, just as it is. What a relief when we can throw all our truths along with all our falsehoods into the cosmic wastebasket! But how? How can one accomplish such a purging feat? How to reach the pure, unencumbered little cell without bringing down the whole structure we have built over it, and crushing the little cell underneath? This is where we truly need a divine magician. One even suspects, with Aristophanes, Molière, and Sri Aurobindo, that that magician would have to be somewhat of a humorist.

But let's be serious, for the moment. There are also those ominous bombs we are stockpiling like moles in their tunnels.

> **66.219** — Men don't know (they should know, but they don't) that things have a consciousness and a power of manifestation of their own, and that the existence of all those destructive devices impel their utilization, so that even if men don't want to use them, a force stronger than they are would push them to use them.

"Things" have consciousness, whether they are bombs, cells, or atoms. The truth is, the entire universe is consciousness and matter is consciousness—precisely what we are not. We mistake intelligence for consciousness, and that is why we see nothing of the universe as it is; we live in an idea of the universe. An explosive idea? Which of the two will finally prevail? That idea or the consciousness in matter? It is like a race between the two, and this is where we are in the race. In her body, Mother was racing between the force of destruction and the other one.

She left, apparently.

And so did Sri Aurobindo, for the same reasons.

Is the combined resistance and negation of the little spiritualists and little materialists greater and stronger than the evolutionary drive? That we are going to change course is beyond a doubt. Those who still see in Sri Aurobindo and Mother great "sages" or "saints" or philosophers, or what have you, are non-evolutionary dimwits, slowpokes from the Ice Age of spirituality. Mother and Sri Aurobindo did not come to preach or reveal; they came to DO. And they did what they had come to do. "The thing is done." They came to untangle and set free a group of cells in a small corner of matter, a small portion of human cellular substance—a group of cells as they are without their coatings and evolutionary crust—despite, or with the help of, all obstacles. Their bodies were a laboratory of evolution. What they did is an evolutionary operation.

Has evolution ever been known to fail?

It is the one thing that never fails, the most infallible thing in the world. Gospels may fail, but not the cell. Once it has something in "mind," or in its program, it will stick to it no matter what, until the next evolutionary upsetter comes along.

Mother and Sri Aurobindo are great upsetters. We just have to take a look around us.

But as usual, we see nothing except slogans and millions of radios and televisions blaring out lie-truths or truth-lies all over the world, and no one can make sense of anything, except that our foundations are shaky.

65.203 — They feel the ground is no longer solid beneath them. It's shaky. They don't feel comfortable.

63.189 — It is impossible for any change to take place, even in a single element or in a single point of the earth consciousness, without involving the entire earth in this change. It's inevitable. Everything is tightly interwoven. And a vibration in one place necessarily has worldwide—I am not saying universal; I am saying worldwide—consequences.

And Sri Aurobindo:

The stone lying inert upon the sands, which is kicked away in an idle moment, has been producing its effects upon the hemispheres.[1]

If our radios and televisions have such effects and can cause panic from Moscow to New York in three minutes, then what do we know of the devastating impact of a little bit of matter

1. *Thoughts & Aphorisms,* XVII, 92.

that has unexpectedly pulled off the stunning coup d'état of overthrowing the government of the mind? This is what we cannot measure, but which is showing its full measure every day in front of our eyes. The mental rule of the world is shaking and falling into incoherence. They are all making speeches, but the earth is shaking. Matter is shaking. It might be time for the world to awaken to the reality of the phenomenon before all the little hats of presidents, bishops, biologists, yogis, and ayatollahs come flying off before our bewildered eyes.

This is not a "spiritual" turning point in the world; we are not about to change our way of thinking; we are about to change worlds, like the teleosts in their dried-up waterhole. All our communism and Marxism are as ludicrous as our capitalism or our gospels, and so are all possible "-isms"; we are at an evolutionary turning point. The battlefield is the body, in the cells. The way to change the world must be found in the body, in the cells. *That* is what is changing, and nothing else: "All the bodies! All the bodies," she said. Everything else is brain flatulence.

So here we are, at this strange crossroads between apocalypse and biology.

Suddenly the problem became crystal clear. It was in 1969, during an experience quoted earlier, but not in its entirety. And the ending is interesting. Again:

> **69.315** — And what is this creation after all? Separateness, meanness, cruelty—the thirst to harm—and suffering; and then disease, decay, and death—destruction. All that is part of the same thing. But what I experienced was the IRREALITY of those things, as if we had entered an irreal falsehood, and everything disappears when we get out of it; it DOESN'T EXIST, it no longer is! That's what is so frightening! All these things that are so real, so concrete, so frightening for us do not exist. We've just entered a falsehood. Why? How? What? . . . Never, ever, not once in all

its life, had this body felt such total and profound pain as it did that day, oh! Then at the end of it: bliss. And, puff, everything disappeared, as if all this, all these awful things, didn't exist. And all the means—which we could call artificial, including Nirvana—all the means of escaping are worthless. From the fool who kills himself to put an "end" to his life—that's the most foolish of all—to Nirvana, which gives the impression of escaping. All are worthless. They are at different levels, but they are worthless. And after all that, just when you feel you are stuck in eternal hell, suddenly . . . suddenly, a state of consciousness in which everything is light, splendor, beauty, happiness, kindness. It is impossible to express. "Here I am"—it comes—then zap! Gone. Is that it? Is that the key? I don't know. But salvation is physical; not at all mental, but PHYSICAL. I mean, it isn't found in flight; it's HERE. And it isn't veiled or hidden or anything; it's HERE. What is it, within the whole, that prevents us from experiencing "that"? And why? I don't know. It's here. It is HERE. And all the rest, including death and everything, becomes truly a falsehood, meaning something that doesn't exist. . . .

Then Mother added:

But one cannot get out all alone.

64.283 — This is not being done for ONE body, but for the earth.

And this is truly the heart of the problem.

It is not a matter of sorting out the "just" from the "unjust," but of getting out completely and all together of a common glass bowl of irreality where all our marvels and truths, all our monstrosities and falsehoods vanish into something *else*, which changes everything. "A little nothing that changes everything," she said.

The apocalypse is in the core of the cell.

*
* *

The fact is, we have no time left.

One might have hoped that, given time, a few evolutionary heroes would understand the process, go down into the body by forcing their way through the layers, and free the cells from their atavistic and Newtonian spell; then the process would spread, just as the mental process must have spread among the great apes. But it is already spreading, at a dizzying speed! And time is something we no longer have. Dark throngs are rising up for the assault. The earth is crying out in despair. Millions of men are preparing to swoop down upon the earth. A fiery hurricane is whirling over Asia. Do we really believe that the lovely walls of our intellectual glass house will protect us from this senseless, fiery flood? Has anyone ever seen a crazed mob? A colossal, surreptitious contagion is seeping through our anthill-size walls, but which is it? The contagion of the new life or that of imminent death? Behind its cushioned walls, America is wrapped up in electronics and playing with fire. Behind its ramparts, the Kremlin is cornered and scared. A cruel and soulless yellow cat is watching the game, spinning its web and waiting for its hour, while a corrupt India, once the cradle of light, harbors devils in its ashrams and remains the invisible pivot of the battle. For India is the heart of the earth—weighed down, mired, but the heart just the same. Which of the two will win the senseless race—the new life or always the old death? Clearly, it is no longer a question of decades; we have only a few years . . . or perhaps months. It is at our doorstep.

But this new life and that death seem to be so tightly interwoven, not only within each continent but within each nation, each community, each family, within the consciousness of each human being, that one cannot imagine how it would be possible to uproot one without the other. All the headlines are screaming and lying; truth is entangled with falsehood in a single bundle of deceit; falsehood harbors a flicker of truth on which it feeds and which it uses for protection. One cannot touch one point without touching the whole.

And this is, really, where the impossible miracle becomes the only possible miracle: within the cell, within the very body of the earth.

Four of Mother's comments, when considered together, seem to give the clue.

> **66.263** — The ordinary consciousness lives in a state of constant fidgeting; it's a frightening realization! As long as you' don't realize it, it seems perfectly natural, but when you realize it, you wonder how people don't become crazy; it's a grace! It's like a tiny microscopic trepidation. Oh, how horrible! . . .

Exactly the description of the "web" of the physical mind, on the other side of which lies the miracle, all possible "miracles," or more precisely, not a "miracle," but the end of our scientific and mortal falsehood: the unknown "natural" state. Mother added this illuminating comment:

> And it is the same for everything: for world events, natural cataclysms, and mankind; for earthquakes, tidal waves, and volcanic eruptions; for floods and wars and revolutions and people who take their own lives without even knowing why; everywhere, they are all pushed by something. Behind that "fidgeting" is a will for disorder that is trying to prevent the establishment of harmony. It is there in the individual, in the collectivity, and in nature.

So we begin to see that that "web" is not just about individual cells; it covers the entire human world. A constant microscopic trepidation enveloping the world within its net.

Then in 1969:

> **69.105** — The number of suggestions one could call "defeatist"

in the earth's atmosphere is just overwhelming! So much so that it is surprising everything isn't crushed to death. Everyone, constantly, is creating catastrophes; expecting the worst, seeing the worst, considering only the worst. And down to the smallest things, you know (the body observes everything). When people react harmoniously, everything goes well; when the reaction is what I now call defeatist, the person picks up an object and he drops it. That happens all the time, without any reason whatsoever; it's the presence of that defeatist consciousness. And I have seen this: All the wills or vibrations (in the end, everything boils down to different qualities of vibration) that bring about little troubles to the greatest catastrophes all have that same quality!

And in 1971, unexpectedly, I was left dumbfounded:

71.78 — I have a curious impression of a sort of web—a web with meshes . . . very loose, not tightly woven meshes—connecting all events; and if you have power over one of these webs, a whole range of circumstances can change, although they are apparently unrelated to one another. And I feel it is something enveloping the earth. AND IT ISN'T MENTAL. These are interdependent circumstances, linked to one another in a way that is invisible to the external eye and is outside mental logic. If you are conscious—truly conscious of this—that is how you can change circumstances.

(Question:) And you feel you have power over one of these webs?

No, not in that way. It is because I was actually acting on one of these webs that I became aware of them. . . . If one had the power to replace one of these webs with another, one could change all things that way. It is inexpressible.

Which web are you acting on at the moment?

But I have no idea! These are webs going around the earth. . . .

And this is what left me dumbfounded:

I can see one of them. Why, all the tiny circumstances of life are

> on this thing! Looking at it as I am looking at it now, I see that it covers the whole country [India], and not only this country, but the whole earth.

One drops an object here, and something is set in motion in Kamchatka or Washington. And what microscopic vibration (or even gigantic one) in Spitsbergen or on Bourbon Street has induced my misstep all the way here? Everything is interconnected! It is frightening. And it is not mental. Then what is it? Every cell and every atom of the earth within the same continuous body. But then, if you touch a little cell here, if you make a tiny hole in this particular mesh, in this microscopic "personal" web—but nothing is "personal"! Nothing is individual. One cannot puncture one point without puncturing the whole thing! This is what Mother and Sri Aurobindo did—they sowed an irrepressible contagion. Then we see that the problem takes on an unexpected dimension, since the microscopic individual that we are assumes disproportionate importance, on a par with anything on the planet: an earthquake or a beautiful movement of the soul that creates a sudden breath of fresh air in the black brew of the world are equally important. Everything is equal. Only the quality of the vibration matters—dark or light, sunny and happy or defeatist.

And again, this is *not* poetry.

One day in 1967, Mother suddenly emerged from a long concentration or contemplation and began to speak in English, as if it were Sri Aurobindo speaking (this often happened). She spoke in her slow, tiny crystal voice, and I did not understand anything. But now it is clear:

> **67.251** — After some time, I will be able to say *(long silence)* what is meant exactly by the irreality of this apparent matter. I feel, I truly feel on the verge of finding a key—a key or a "knack," a procedure (I don't know how to say it; all these words are a

> vulgarization), but something that could cause one to bring about a frightful disaster in a split second . . . if one were to possess it without being totally on the right side. What disaster? I don't know—something like the dissolution of the world.

The rending of the web? The sudden landing on the true side of the earth? Perhaps quite a startling landing.

But again, this is not science fiction, either.

One year later, it was May 1968.[1] The moment Mother heard about it, she knew: "This is not a strike; it's a revolution . . ." apparently abortive, drowned in the old habit and all the old political game-playing. Still, *something* was there, which may well have been the rehearsal of a more comprehensive, worldwide event yet to come. One could say a "collective, and momentary, tear" in the web. On May 22, Mother said:

> **68.225** — There is a very strong feeling in the consciousness—very strong—that the time has come. There are immense periods of time when things are being prepared. The past is exhausted and the future is prepared, and these are immense periods, dull and colorless, when things go on repeating themselves, on and on, and it seems it is always going to have to be that way. But then, suddenly, between two such periods, the change takes place. Just like the time when man appeared on earth. Now it's something else, another being.

And Mother was looking at those students, the youth of the earth:

> The police represent the defense of the past. But if MILLIONS—not thousands, but millions—got together and occupied [the universities], absolutely peacefully (simply gather together and sit there), then that would have power. But there should be no violence; the moment one indulges in violence, it's the return to the past, the opening to all conflicts. It should be an occupation

1. The student protest throughout the world, and in particular in France.

> by the masses, but masses made ALL-POWERFUL BY THEIR IMMOBILITY, imposing their will through their sheer number. This is clearly—not in the details, but in the general direction of the movement [of May 1968]—the will to be done with the past, to open the door to the future. It's something like a disgust with stagnation. A thirst for "something" ahead, which seems more luminous and better. Indeed, There IS something. It isn't just imagination: There IS something. That's the beauty of it: there IS something. There is an answer. There is a Force seeking to express itself.

The web can be torn—it is ready for it—if enough millions of little vibrations of hope want to cry out, to cry NO to all this irreal falsehood.

And then we approach the fairy tale.

But a very rational fairy tale, maybe the supreme rationality of this world.

Breaking through the web is not a vain fantasy. It is something that all of us, or many of us, especially children, may have experienced without knowing it. They land on the flintstones at Fontainebleau unhurt, without a scratch, *as if nothing had happened.* Actually, nothing *has* happened. Those moments we call spells of heroism or somnambulism—or any -ism—when suddenly everything feels light and the body dances as if one with surrounding matter, and the gaze is as clear as a flame and we go through anything: fire, bullets, accidents, death. Nothing can touch us. We are invulnerable. We are triumphant and light. We do not even think about it; it is simple, so simple, self-evident, without fuss. The lungs are filled with the sweet scent of the springtime air, everything is easy and malleable. All we have to do is say: "I want," and there it is; we are right in the miracle. False matter falls away, we are one with the great wind that gently carries the worlds. We have all known such moments. The web loosens. Everything is different.

> **64.253** — This is my experience of these last few days, with a vision and a conviction, a conviction based on experience: the two vibrations [the trepidating one of falsehood and the light one of "truth"] are constantly, constantly intermingled, and constantly one infiltrates the other. Perhaps the wonderment comes when the amount infiltrated is great enough to be perceptible. But I have the feeling, a very distinct feeling, that this phenomenon is happening all the time. All the time, everywhere, in a minute way, there is a sort of infinitesimal infiltration of truth into falsehood, and under certain conditions that are visible—it's like a luminous expansion; I can't describe it—the amount of infiltration is great enough to give the impression of a miracle [perhaps what happened in May 1968]. But otherwise, it is something happening all the time, all the time, ceaselessly, throughout the world.

The substitution of vibration.

The miracle of the earth taking the place of its falsehood.

And what if it happened collectively? What if millions, yes, millions of young voices, tired of the old earth of falsehood and its gray columns of humans lining up to receive their degrees in the old way of dying—what if those clear little voices suddenly let their hearts melt, let their chests fill with spring air, and shouted, "ENOUGH! we've had ENOUGH of all this!"

All those cells suddenly freed from their hypnotic state.

"When? When?" asks the voice of the earth.

> **33.1210** — I think it will happen when a great enough number of conscious people feel that there is absolutely no other way. Everything that has been, and still is now, has to appear as an absurdity that can no longer continue; then it will be able to happen, but not before. Despite everything, a time will come when it will happen; a time will come when the movement will shift toward a new reality. There has been a MOMENT. There has

been a moment when the mental being was able to manifest on the earth. There will be a MOMENT when the human consciousness reaches a state enabling the supramental consciousness to enter that human consciousness and manifest. It doesn't stretch like a rubberband, you see; there comes a moment when it happens—it can happen in a flash.

We will drop everything: our pens, our laws, our science, our caged-in future. The world will swell with an immense laughter, and we will be there!

But why not now?

The already dead will simply drop dead.

An apocalypse, yes, a smiling one.

Fatal for the dead and light for the ever-living.

A fairy tale within the cells of the earth.

Land's End

February 15, 1980